# HISTORY OF SCIENCE

John Priestley

MA (Oxford)
MSc (Liverpool)

# Copyright page

Published by Lulu Books

ISBN 978-1-326-20367-2

© Copyright 2015

First Edition

**By the same author:**

*Tyke on a Bike* – the canals of northern Britain, as viewed by a Yorkshireman
*History of the British Isles to 1714 AD*
*History of the British Isles 1714-2010*
*History of England* (also published as *History of Britain*)
*Oracle e-Business Consultancy Handbook* – Essays, Hints and SQL scripts for system users in Manufacturing, Supply Chain and Finance
*Armchair Geology* – including a stratigraphical history of the British Isles, also published as *The Armchair Geologiss*
*Jeeves and Wooster Short Stories*
*Geography of China*

List of illustrations ...................................................................................7
Chapter 1 – Early Science ..........................................................................9
Chapter 2 – Renaissance Scientists ..........................................................17
Chapter 3 – The Scientific Revolution of the Seventeenth Century ......27
Chapter 4 – Into the Eighteenth Century ................................................44
Chapter 5 – The Emergence of Chemistry ...............................................51
Chapter 6 – Later Georgian Science ........................................................62
Chapter 7 – The First Geologists .............................................................77
Chapter 8 – Life Sciences in the Nineteenth Century .............................86
Chapter 9 – Victorian Science ..................................................................97
Chapter 10 – The later Nineteenth Century ..........................................108
Chapter 11 – Albert Einstein ..................................................................124
Chapter 12 – Early Twentieth Century Chemistry and Physics ..........133
Chapter 13 – The Development of Genetics ..........................................149
Chapter 14 – Earth Sciences 1890-1970 ................................................170
Chapter 15 – Twentieth-Century Astronomy .......................................186
Chapter 16 – Computer Science .............................................................200
Chapter 17 – Atmospheric Chemistry ...................................................209
Chapter 18 – Recent Geology .................................................................214
BIBLIOGRAPHY ...................................................................................219

**List of illustrations**

Hippocrates
Aristotle
Retrograde motion of the planets
Nicholas Copernicus
Conjunction of the planets
Elliptical orbit of the Earth
Isaac Newton
White light as split by a prism
Experiment with a Bird in an Air-pump
Benjamin Franklin
Young's double-slit light experiment
Geological unconformity
Ichthyosaur
Mont Blanc
Charles Darwin
Gregor Mendel
The inheritance of pea colour
Valency
Structure of the carbon atom
Faraday's experiment creating motion from electromagnetism
Structure of methane
The hydrogen bond
Alpha, beta and gamma rays
Harmonic and interfering waves
Fruit fly inheritance
Mammoth Hot Springs
Frontal weather systems
Precession of the Equinoxes
The Mid-Atlantic Ridge in Iceland
Subduction
Transform Faults
Parallax measurement
Ichthyostega
Dimetrodon

## Chapter 1 – Early Science

This book will not dwell for long on ancient and medieval science, for the simple reason that most of it turned out to be wrong. Nevertheless mention must be made of the first scientists, who were of course the ancient Greeks. After the great flowering of the civilisation of Athens in the fifth century BC, Greek culture spread throughout the eastern Mediterranean, and it was sometimes these new centres which supplied the earliest scientists. The most important of these centres was founded by Alexander the Great himself in 331 BC, the eponymous Alexandria in Egypt. Despite its later political eclipse by the Roman Empire, Greek culture continued to flourish for hundreds of years over a very wide area.

The title of first scientist is generally accorded to Thales of Miletus, said to have predicted the eclipse of the Sun which took place in 585 BC. Certainly one of the first mathematicians must have been Pythagoras (570-495 BC), whose theorem of right-angled triangles is still valid today. We all learnt it at school!

The square of the hypotenuse is equal to the sum of the squares on the other two sides.

This man also derived a value for Pi, at the same time concluding that this value is what we call an irrational number – it cannot be stated exactly. Another name which survives from the early years of Greek civilisation is Hippocrates (460-370 BC), the father figure of medicine, whose name we know from the Hippocratic oath taken by doctors. (Note that the lifespans of the ancients are usually approximations, and indeed it does not seem likely that Hippocrates lived a full ninety years!). The ancients had also glimpsed the possibility of the atomic structure of all matter. Democritus (also 460-370 BC) and his mentor Leucippus devised an atomic theory which stated that all matter is made of tiny particles or atoms which are indivisible and indestructible.

Many ideas which survived for centuries came from Aristotle, who lived in the fourth century BC. His model of the universe with the Earth at its centre was to survive until 1510, and as far as the Catholic church is concerned, until 1835! Democritus was known to Aristotle, but Aristotle rejected his ideas.

Early chemistry also hailed back to Aristotle, who apparently made the mystifying (to modern eyes) but long-enduring claim that everything is made of a mixture of fire, earth, water and air – surely Aristotle must have know there was more than one kind of earth! It makes a little more sense if we substitute the word "energy" for "fire". In the case of water, Aristotle made an understandable mistake, because nobody, on the bases of their common sense and everyday powers of observation, could possible guess that water is in fact not an element, but is composed of two inflammable gases – hydrogen and oxygen! For that matter – science again defying common sense – how could anyone realise that a substance like common salt is in fact made by chemical bonding between a dangerous and highly reactive metal (sodium) and a poisonous gas (chlorine)? Aristotle then added a fifth element, the ether, in which sat the bodies of the heavens. We derive the words "ethereal" and "quintessence" from this root.

If Aristotle did not acknowledge different kinds of earth, then others of his era did, and in fact eleven modern elements were already known to the ancient Greeks: antimony, carbon, copper, gold, iron, lead, mercury, platinum, silver, sulfur and tin. The ancient Egyptians before them recognised seven elements, and associated them with the seven known heavenly bodies, a usage which lingers on to this day: gold for the Sun, silver for the Moon, and mercury for Mercury!

The geometry of Euclid of Alexandria, still taught in schools, dates from around 300 BC. As early as the third century BC, one Aristarchus of Samos described a model of the heavens with the Sun, rather than the Earth, at its centre. Other famous scientists were Archimedes (287-212 BC), who came from Syracuse in Sicily, and the astronomer Ptolemy (90-168 AD), who lived in Alexandria. A contemporary of Archimedes was Eratosthenes (276-195 BC), the librarian from Alexandria who measured the circumference of the Earth! From the Greek heartland itself – Pergamon, in Asia Minor – came the greatest name of ancient medicine, Galen (129-200 AD).

Of these scientists and mathematicians, Archimedes established a principle which has survived intact – that a floating body displaces its own weight in water. This apparently occurred to him whilst floating in his own bath, causing him to cry "Eureka"! – "I have found it!" A fully immersed body displaces its own volume in water, but if it floats, this changes to its own weight. This has some important consequences today in the area of global warming. If the Arctic ice sheets melt, this will have no effect on worldwide sea levels, since they already displace their own weight in water. If, on the other hand, the Greenland ice sheets were to melt, then that would cause a dramatic rise in sea levels, because it would represent new water flowing into the oceans which is currently locked up on land. Archimedes, also famous for inventing the spiral water pump or screw which carries his name, had the misfortune to be murdered by Roman soldiers invading Syracuse. Despite orders that his life should be spared, he was killed before he was recognised; nevertheless he was by that time about 75 years old.

In his way even more remarkable was Eratosthenes. Born in what is now Libya, he became a librarian in the Great Library of Alexandria. By his time, the concepts of the equator and the Tropics of Cancer and Capricorn were understood, the tropical meridians being the points at which the Sun turned back in its seasonal movements. Eratosthenes understood that the Egyptian city of Syene (modern Aswan) lies on the Tropic of Cancer, as at noon at the summer solstice, the Sun is directly overhead there (casting no shadow as it shone down a well). At the summer solstice in his home town of Alexandria, he found that the Sun is at an angle of 1/50 of a whole circle, or 7° 12". He did this by observing the shadow of the sunlight as it entered a well. Assuming that the Earth is a full circle of 360 degrees and that Alexandria lies directly north of Syene (both correct), then all he needed to know was the distance between Alexandria and Syene, and he could work out the circumference of the whole Earth. The distance to Syene was already known, and he corroborated this by enquiring how long it took a camel to get there. From this information he calculated that circumference of the Earth is (in modern measurements) 39,630 kilometres (24,600 miles) – within 2% of the modern figure.

Eratosthenes also correctly calculated the distance between the Earth and the Sun; the tilt of the Earth's axis (23.5°); and invented the system of latitude and longitude – all without leaving his home in Alexandria – quite an achievement then! It is said that in Medieval times, some people thought that the Earth was flat, but this clearly did not apply at all to Eratosthenes. One big clue was that it was perfectly obvious that both the sun and the moon were globes, so probably the earth was as well.

The Romans, who came to rule the lands of Ptolemy and Galen, though great builders who produced monuments unrivalled for a thousand years or more, made few advances in science or mathematics. In any event, most of the works of the Greeks were lost to western Europe for centuries after the fall of the western Roman Empire in the fifth century AD, though they survived in the eastern Mediterranean, where political mastery passed back to the Greeks in the Byzantine Empire. It was from here that many were later translated into Arabic.

The Arabs made some progress with chemistry. The best-known of them is Geber or Jabir, who lived in Baghdad in the eighth century. He managed to accumulate a basic set of ingredients. One of these was sal ammoniac (ammonium chloride), which is deposited around the mouths of volcanic vents. He also distilled vinegar to make a strong acetic acid, and managed to prepare a weak solution of nitric acid. Another chemist, or more specifically pharmacist – and in the later Middle Ages, a very famous one – was Avicenna (980-1037). He was a Persian from near Bukhara in what is now Uzbekistan. His book *Canon of Medicine*, a pharmacopoeia which detailed the medicinal properties of chemicals and plants, became a standard text in western universities.

The next chemist of note was an anonymous European of the thirteenth century, known as "False Geber" as he signed his works "Geber". It was he who discovered how to prepare sulphuric acid, about 1300, and then known as oil of vitriol – and, as every schoolboy knows, you can't get far in chemistry without sulphuric acid! False Geber also described how to make concentrated nitric acid, known as aqua fortis. These liquids are highly reactive and in sufficient strength can dissolve metals to create salts such as sulphates and nitrates of copper. Their discovery was a major step forward for chemistry, albeit one which had little immediate impact.

It was also in this period that the first useful new metal to be discovered in thousands of years was first isolated – zinc. It was extracted from its ore, calamine, a zinc oxide famously used in medicinal preparations, in India in the thirteenth century. However it was not until the eighteenth century that exploitable ores were found for zinc in Europe.

Until as late as the seventeenth century, there was then little chemistry as such, but only alchemy to search for the philosopher's stone, thought to be a catalyst which would turn base metal into gold. Needless to say, no one ever found this, despite years of searching by some otherwise respectable people, including Isaac Newton, and the wastage of an unconscionable quantity of eggs (which have, of course, gold yolks). Other hopeful materials were sand and urine (both yellow). There was a Chinese version of alchemy, but the objective was different. The Chinese sought an elixir which would confer eternal youth.

In terms of science (if not, say, of architecture), there were no advances in western Europe until the Renaissance, the rediscovery of the knowledge of the ancient Greeks in the later Middle Ages. This was greatly encouraged by the fall of Constantinople to the Turks in 1453 AD, as this caused a diaspora of learned men to the cities of the west. From this time onwards, old texts became increasingly available in one form or another. One of these was called *De Rerum Natura* (*On the Nature of Things*), a Latin poem from the first century BC by Lucretius, a single copy of which survived into the Renaissance. Amongst other things, it contained the idea that all matter ultimately consists of atoms.

Educated men looked at the ancient texts with new eyes, and came quickly to realise that improvements could be made. Even the Romans had been in awe of Greek learning, but the new thinkers of the late Middle Ages were not. This rethinking affected not only science, but also religion. The first significant breakthrough in science – the heliocentric solar system of Copernicus– came by 1510, and in religion – the permanent establishment of Protestantism by Martin Luther in Germany – in 1517. Beguiled for so long by those twin impostors, alchemy and astrology, the scientists of the late Middle Ages were about to break the codes of a new world of knowledge.

The first of all the sciences was one based on direct observations of nature, astronomy. Up until 1510, most astronomy derived from Ptolemy and his great work, known generally as the *Almagest* ("The Greatest"). There were only five known planets – Mercury, Venus, Mars, Jupiter and Saturn. Joining them in the heavens were the Sun, the Moon and the "fixed" stars. It was held that all these bodies circled in the sky around the Earth, each set in its own "crystal sphere" – crystal in this case meaning clear, rather than made of glass; and according to Aristotle, embedded in the ether. In fact some ancient Greeks had already come up with the idea of a heliocentric system, with the sun, not the earth, at its centre. The fact that Ptolemy ignored this concept was a serious error which was to mislead thinkers and navigators for a thousand years.

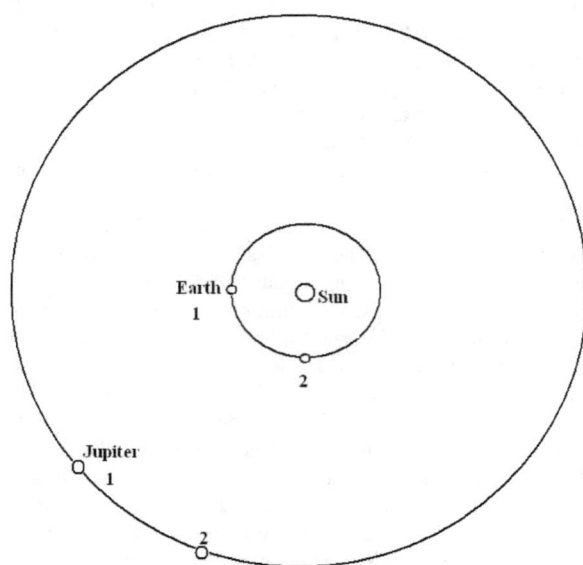

Jupiter at first appears ahead of the Earth in the sky, but at the second observation point it has gone retrograde, or fallen behind.

In fact there were problems – fully realised by Ptolemy – which could not be explained on the simple model of the earth at the centre of the universe. The first of these was that instead of simply going round the Earth in a forward direction, the planets could appear to move backwards; something known as retrograde motion. This is quite easily explained on the modern scheme, where we know that the Sun lies at the centre of the solar system. The Earth orbits the Sun, and beyond that, much further away, so does Jupiter. As the Earth overtakes Jupiter on its short inner circuit, then instead of moving ahead in its orbit, Jupiter falls behind. For this reason, Ptolemy and the ancients invented the epicycle, a device where each planet had its own mini-orbit within its own crystal sphere, like a wheel within a wheel. This solved the main problem, but the orbits calculated in this way were still not consistent with the observations. So it was that a second abstract construction was invented, known as an equant. This was an offset from the Earth around which each crystal sphere revolved – the Earth was near but not quite at the centre of the universe; in fact the universe had many centres on this scheme.

There were also problems with the orbit of the moon. In order to account for the changes in speed with which the Moon appears to cross the night sky, Ptolemy's scheme required that the Moon should be significantly nearer to the Earth at some times of the month than at others (and so its size should also change noticeably, and by a calculable amount). However, the size of the Moon did not change in this way; everyone involved was aware of this, but simply brushed aside this inconvenient truth, as it were to await a better explanation in later times.

Finally, there was one issue which even the thinkers of the Renaissance were unable to resolve – the mechanism which lay behind the movement of the heavenly bodies. Until the time of Newton, no one had any idea how it worked, and, as usual in these cases, ascribed it all to the hand of God.

It is a well-known aspect of science that messy theories are unlikely to prove correct, and Ptolemy's was a messy theory, but it was all there was for almost 1400 years after the publication of the *Almagest* around 147 AD. Great theories are characterized by elegance or simplicity – after all, what could be simpler then this?

$E = mc^2$

(Einstein's fundamental equation of relativity.) It is also a notable feature of science that common sense is not necessarily a very good guide. The great virtue of the Ptolemaic theory was that it fitted in with the seemingly obvious fact that the Earth stood still while the rest of the universe spun around it. However there were grave doubts about the theory, because if even a tiny discrepancy is found in the observations, this is likely to mean that a theory is wrong.

Ptolemy also produced a second great book, *Geography*, an annotated atlas of maps. He set out the principles of lines of latitude and longitude here, and defined useful concepts in map-making including the idea of minutes and seconds of degrees. However, once again, he made a great error, by rejecting Eratosthenes' estimate of the circumference of the earth, instead accepting another which made the earth seem only three-quarters of its actual size. He also assumed that the known world covered 180 degrees of longitude, from the Canaries in the west to the eastern tip of Asia in the east. He seems to have had no evidence to support this, as Asia only extends this far in the frozen north-east of Siberia; at navigable latitudes, 180 degrees from the Canaries stretches to the middle of the Pacific. These errors, once again, were to persist for a thousand or more years, seriously misleading the early global navigators, including Columbus.

## Chapter 2 – Renaissance Scientists

The man who started the heavenly revolution did not mean to do so. He was a German who went under the Latinized name of Regiomontanus. In his book the *Epitome*, he sought to summarize the *Almagest*. He also added more observational data, and provided a commentary. It was in this commentary that he pinpointed the problem of the orbit of the Moon and the way in which its apparent size does not change in the way that Ptolemy requires. When the books was published in 1496, twenty years after the death of Regiomontanus, this crucial point was noted by a young man of twenty-three called Nicholas Copernicus. Had the book been published during the lifetime of its author, it is likely that someone other than Copernicus would have achieved a place in the scientific pantheon.

Copernicus (1473-1543) came from the town of Torun, on the River Vistula in the north of Poland. There remains some doubt as to whether his native language was German or Polish, because for centuries this area, known as Old Prussia, was of mixed German and Polish settlement. His father was a wealthy merchant, but he died when Nicholas was still a boy of ten or eleven. Henceforth his patron would be his maternal uncle, Lucas Watzenrode, who in due course became the Prince Bishop of Varmia, in this region, at a time when bishops were both rich and powerful men. As a result, Copernicus received a very expensive education, much of it in Italy. He emerged from this as a physician. Meanwhile in a typical piece of graft from the Middle Ages, his uncle saw to it that he became one of the twenty-four canons of Frauenberg (Polish Frombork) Cathedral. The practice of nepotism (the appointment of nephews – in practice often illegitimate sons – of bishops or the Pope himself to church sinecures) was one of the besetting sins of the Catholic Church, later frowned upon as such by Luther and the new Protestants. At first this job was a pure sinecure, with no duties attached, though on his return to Poland Copernicus did become involved in the administration of the vast estates of the bishopric.

Nevertheless Copernicus set to thinking about the configuration of the heavens, one of the intellectual hot potatoes of the day. This in fact did not involve much else but thinking, as he was no great observer of the heavens. He particularly disliked the idea of equants, as these offsets meant that the system as it stood had no single centre. The result of this "thought experiment" was the heliocentric theory of what became known as the solar system. It is known that Copernicus had reached his conclusions by 1510 as he circulated his ideas in letters to friends and acquaintances. Note that there is nothing wrong with thought experiments – effectively speculation on existing data – and indeed these were to achieve great things for Albert Einstein in his *annus mirabilis* of 1905, when he published a string of new theories, including special relativity.

Nicholas Copernicus

Copernicus placed the Sun at the centre of the system, with the planets including the Earth orbiting around it, newly lined up in the correct order – Mercury, Venus, Earth, Mars, Jupiter and Saturn, with the fixed stars beyond that, and the Moon with an independent orbit round the Earth. With this system, many of the discrepancies in the Ptolemaic system disappeared at once. There was no need for equants, though a form of the epicycles remained, as the new model still could not explain why the planets appeared to speed up and slow down in their orbits (the reason for this was that the orbits are not perfect circles, but ellipses). However another great puzzle disappeared immediately. This was the known fact that Venus and Mercury could only be observed at dawn and dusk, and never at night. The reason for this became clear – their orbits lie between the Earth and the Sun, so at night, when the Earth faces away from the Sun, they cannot be seen.

Copernicus was loathe to publish his new theory for fear of upsetting the notoriously conservative Catholic authorities. He was also aware that the new theory raised new problems. If the Earth span round at the speed required, why wasn't there a constant gale caused by this rapid motion? Again, if the Sun lay at the centre, why didn't everything else simply fall into it? These were real problems, but they do demonstrate that a good theory – and it was a very good theory – doesn't have to have all the answers.

After the death of his uncle, Copernicus was a busy man with both his medicine and his church duties, though he sometimes got into trouble with the new bishop, especially in the matter of his "housekeeper", one Anna Schilling. Towards the end of his long life, however, he was visited by a young German Lutheran academic known as Rheticus, who, in the manner of a modern literary agent, coaxed the book out of him. In 1540 Rheticus himself produced a pamphlet called the *Narratio Prima* ("*First Account*"), summarizing the key features of the new model.

Rheticus then oversaw the production of Copernicus' own book with the publisher, Petreius of Nuremberg. Before the work was completed, however, Rheticus had to leave town in a hurry, as he was accused of making drunk and sexually abusing a young man. He absconded to Leipzig, leaving the work to be overseen by a Lutheran theologian, Andreas Osiander.

As Copernicus lay dying in 1543, the work finally appeared in print as *De Revolutionibus Orbium Coelestium* (*On the Revolutions of the Heavenly Spheres*), a highly misleading title since the heavenly crystal spheres were being abolished by the work itself. Along with the book came a preface stating that the heliocentric system it contained did not actually mean that the Sun was the centre of the universe, rather than the Earth. Instead this was merely a model which would simplify calculations involving the movement of the Moon and the planets. Astronomical calculations – in practice the prediction of eclipses and conjunctions (where two planets line up) – were regarded as of the greatest importance, especially amongst the astronomers who were also astrologers. The preface was undoubtedly the work of Osiander, and it certainly infuriated Rheticus, who tried to have it removed. However

Osiander was merely trying to shield the book from the criticism which was certain to follow from the Catholic church.

The book left some baffling questions unanswered. One of these concerned the fixed stars. As the Earth swings round its orbit, it moves a distance of approximately 300 million kilometers (187.5 million miles) in space (or two astronomical units, or twice the distance between the Sun and the Earth) in a direct line traced across the heavens from midwinter to midsummer. Surely, then, the position of the fixed stars would be at least slightly different, when measured from either end of the orbit? This difference is known as the parallax, the apparent movement of an object when measured from two different places. (You can observe this effect very easily yourself by holding up your index finger at arm's length and looking at it, closing first one eye, then the other. It appears to move with respect to the objects behind it, when in fact it has stayed in the same place.) In fact there are parallax differences, but they are miniscule, and far too small to be observed by any means available in 1543. Copernicus came to the correct conclusion that the fixed stars must be so far away that no parallax could be observed, but others were sceptical.

At first it seemed that Copernicus would have had nothing to fear from the publication of his great work. True, Luther's Protestants were as opposed to the ideas it contained as the Catholics, saying what religious people always said, that Joshua commanded the Sun to stand still (in the Old Testament), and not the Earth. The book was in any case not a popular pamphlet – Rheticus had supplied that – but a highly technical work designed for specialists. It failed to sell out its first edition of 400 copies; but to the people who mattered, the next generation of astronomers, it was an absolutely crucial work. A second edition was printed in Switzerland in 1566. Having been largely ignored by the Catholics in the sixteenth century, *De Revolutionibus* was finally placed on the Pope's list of banned books, the Index, in 1616, where it remained until 1835!

By the second half of the sixteenth century, other names had begun to make a mark. One of these was the Englishman Sir Thomas Digges (1546-95), whose father Leonard invented the telescope. This enabled his son to see more of the myriads of stars in the night sky, and to conclude that there is an infinite number of stars, at varying distances from the Earth. This may seem obvious now, but it was a big move away from the fixed stars of Ptolemy. Digges published his findings in 1576.

Another of the next generation of astronomers was Giordano Bruno (1548-1600), an enthusiastic supporter of Copernicus and a man who came to a famously sticky end. Bruno, at one point a Dominican friar, extended the Copernican system by proposing that the Sun is actually a star, that there are an infinite number of similar stars in the universe, and that these also have solar systems with inhabited planets. Bruno also dabbled in Egyptian religion, to the extent that he was accused of heresy. After refusing all requests to recant, Bruno was burnt at the stake in 1600, accused of the heresy of Arianism (the belief that Jesus was not in fact a god). Later generations came to regard him as a martyr for the new science and the freedom of thought, though the fact is, he was actually executed for old-fashioned heresy.

During this period, there took place an event which was a great blow to the old fixed stars system – the arrival of a brand new star! Usually known as Tycho's Supernova, this burst into the sky in the constellation of Cassiopeia in 1572. (Cassiopeia has the appearance of a lopsided W in the night sky.) The man who made the most accurate observations of it was a young Danish aristocrat called Tycho Brahe (1546-1601), who described the phenomenon in a publication called *De Nova Stella* in 1572 (hence the word "nova").

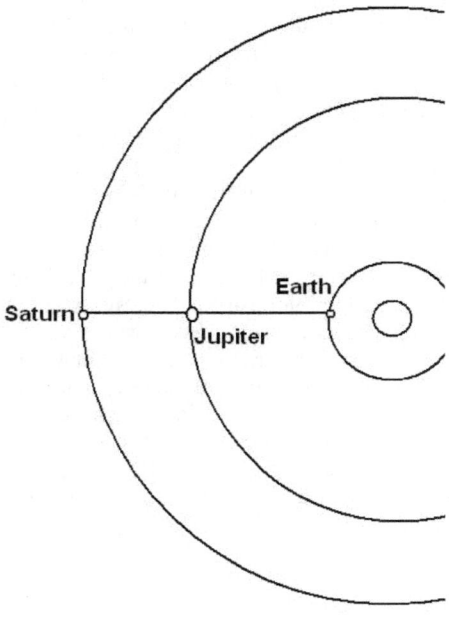

The conjunction of Earth, Jupiter and Saturn

Brahe had something of a reputation as a hothead, having lost part of his nose in a duel with another Dane at the age of 21 (he had an alloy replacement). He had shown an early interest in astronomy and was wealthy enough to design and build his own instruments – giant sextants and quadrants – for observing the stars and planets, and in particular for measuring accurate angles between them. He had quickly come to realise that the measurements of Ptolemy and others were simply not accurate enough to make reliable predictions for celestial events – reliable, that is, down to the exact day. One such event was the conjunction between Saturn and Jupiter, when the two planets appeared to merge, which took place in August 1563. Even the best set of astronomical tables was several days in error in predicting this event, which had been widely anticipated at the time.

Brahe established an international reputation in his twenties, to the extent that he was able to set up what amounted to a research institute known as Uraniborg on the Danish island of Hveen, partly funded by the Danish monarchy. The detailed work involved in the observations was both labour-intensive and time-consuming. For example, it took twelve years to observe the track of both Jupiter and Mars through the constellations accurately.

One event which caught the attention of the sixteenth-century stargazers was the appearance of a comet in 1577. It was apparent from Brahe's observations that the comet (in fact all comets) crossed the orbits of the planets; previously they had been thought to be local phenomena, closer to the Earth than the Moon. This realization was to have an effect on later thinking. The planets all orbit in the same plane around the Sun, and this indicates a common origin. The comets by contrast cross the sky in completely different planes, with wildly eccentric parabolic orbits. This indicates that they do not have the same origin as the planets.

Brahe remained wedded to the Ptolemaic system of the fixed Earth, though he moved some way from Ptolemy's other ideas, including the crystal spheres. He saw no reason for the orbits to be embedded in anything physical, but simply saw the planets as moving unsupported through space. This was a step forward from the old thinking.

Our next astronomer, the German mathematician Johannes Kepler (1571-1630) was certainly not born, like Brahe and Copernicus, with a silver spoon in his mouth; in fact he had to fight every inch of the way of his precarious existence. His father was a mercenary soldier who eventually disappeared, and his mother at one point was tried for witchcraft. Meanwhile Kepler, once established in the world, found himself a Lutheran in lands where most of the power and money belonged to the Catholics. This strife between the Lutherans and Catholics, which had also affected Rheticus, was to break out into open warfare during the lifetime of Kepler. This was the Thirty Years War, which was exactly that – it lasted from 1618 to 1648 and tore the heart out of central Europe, only to terminate with most of the boundaries back where they were in 1618, including the one which divided the northern Protestants from the southern Catholics of Germany.

In 1597 Kepler published a book which noted the fact, observed by Copernicus, that the planets move more slowly in their orbits the further away they are from the Sun. He thought that the planets might be propelled in their orbits by a force which he called "vigour", reaching out from the Sun, which would lessen with distance – not so very far from the truth, as we shall see when we consider Newton and his inverse square law of gravity. He also considered that it would one day be shown that the solar system operated like clockwork, without the need for divine intervention.

In 1600 Kepler eventually arrived at Benatky Castle, not far from Prague, the final home of Brahe and his observatory. Thus it is that the work of the two men became inextricably entwined, because Kepler relied for his subsequent findings on the mass of detailed observations made by Brahe. By 1601 Kepler was assistant to Brahe, but his access to data was limited by the jealousy of the older man. However, that same year, Brahe died aged only 54, and Kepler took his place.

His first task was to try to resolve the riddle of the orbit of Mars, which was observed to move more quickly in one half of its orbit (when it was nearer the Sun) than in the other. At one point he hit on the idea of carrying out his calculations of the orbit of the Earth from the perspective of an observer on Mars. This shows his difficulty – he was trying to establish the orbit of Mars (in fact elliptical) from the Earth, a planet which is itself moving in an elliptical orbit. In this phase of his work he established his "second law", which is that a line drawn from the Sun to a planet moving it its orbit around the Sun sweeps out an equal area in an equal time. That is, when the planet is nearer the Sun and moving relatively quickly, it sweeps out the same area as when it is further away and moving more slowly, because it covers more of the orbit in the same time. As a part of this work he created a formula which showed that a planet's rate of motion is inversely proportional to its distance from the Sun – the planet goes faster nearer the Sun, and slower at greater distances.

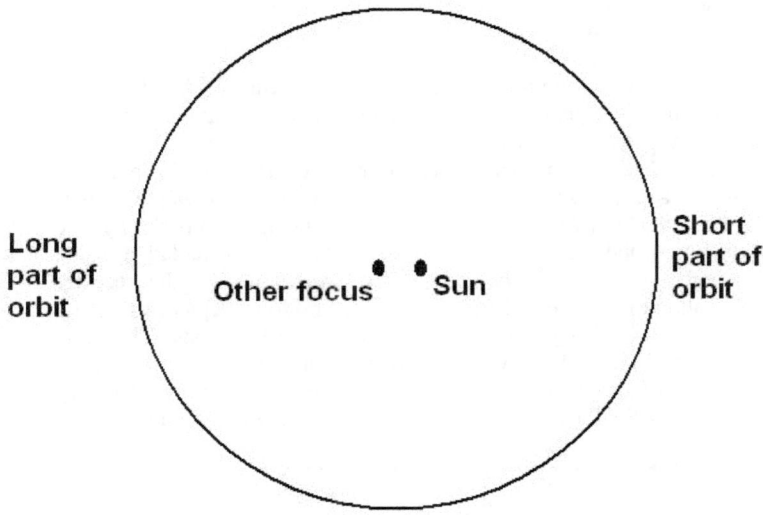

The elliptical orbit with the Sun at one of the foci of the ellipse

By 1605 Kepler had established what is now known as his first law, which is that the planets move in elliptical orbits. Every ellipse has two foci, and the planets orbit around the Sun which lies at one of these foci (in each case, the same one). As a geometrical shape, the ellipse had certainly existed in Greek geometry, but had rather been neglected by mathematicians since then. It was simply assumed that planetary orbits are circular, and they are close enough to circular to deceive the casual observer. However, Brahe was anything but a casual observer, and it was his data which led Kepler to this groundbreaking new concept. So it was that the first and second laws between them finally got rid of any need for equants or epicycles, or for that matter, the hand of God; the solar system did indeed work like clockwork.

Kepler's so-called third law came much later, in a book published in 1618. This states that the cube of the distance of any planet from the Sun is equal to the square of its orbital period (in the case of the Earth, the orbital period is one year). Using the units of the Earth as a basis, Mars has an orbital period of 1.88 years. Its distance from the Sun is 1.52225 Earth distances (it is just over half as far out again).

1.88 squared = 3.53
1.52225 cubed = 3.53

Another way of putting this is that the if the cube of the distance is divided by the square of the orbital period, the resulting number is always the same, for all six planets.

In 1627, after long and weary efforts and many delays, political, religious and military, Kepler finally produced his *Rudolphine Tables*, named after his patron, the Holy Roman (Austrian) Emperor. These were a new and highly accurate set of tables for the calculation of the movement of heavenly bodies. They owed much to the recently publication of tables of logarithms, invented by John Napier of England (1550-1617), which greatly eased the burden of calculations. The value of Kepler's new tables was demonstrated in 1631 when the French astronomer Pierre Gassendi used them to observe the transit of Mercury across the face of the Sun – the first such observation ever made.

\*\*\*\*\*\*\*\*\*\*\*\*\*\*\*\*\*\*\*\*\*\*\*\*\*\*\*\*\*\*\*\*\*\*\*\*\*\*

Between Copernicus and Kepler came the work of an Englishman, William Gilbert (1540-1603), the founding father of the study of magnetism. He established how magnetism works in detail, and what is and what is not magnetic; and he got rid of many of the myths which were current on this subject at that time. He was the first person to recognise that the Earth itself acts as one giant magnet.

# Chapter 3 – The Scientific Revolution of the Seventeenth Century

Our next astronomer, a man often considered the first true scientist (as Kepler was obliged to supplement his income by casting horoscopes) is the Italian Galileo Galilei (1564-1642). Galileo understood that "thought experiments" as undertaken by the ancients whilst strolling about (hence called the "peripatetics" ) were not good enough (though sometimes, they can be!). Having arrived at a hypothesis, Galileo then set about testing it by designing experiments, in the manner of a modern scientist. For example the commonly held assertion that a heavier stone will fall more quickly than a lighter one, dating all the way back to Aristotle, could never have been tested, and when it was, it was easily disproved. That Galileo did this by dropping stones from the leaning tower of Pisa is almost certainly apocryphal!

Here is another man who was often short of money, having spurned the career in medicine which his father had selected on his behalf. As soon as he was introduced to real mathematics (as opposed to arithmetic) he lost all interest in medicine. He became a professor at the university of Padua, then later Pisa, and would probably have lived all his life in obscurity but for an invention which was made in his middle age, the telescope. However his earlier life is surrounded in myths, one of them mentioned above. Another is that he mastered the mechanics of the pendulum whilst watching one ticking backwards and forwards when bored during a church service, timing the swings against his own pulse. He established that the swing of the pendulum depends on its length, and not on its weight, or the length of the arc through which it swings.

He is also noted for rolling balls down inclined planes, to establish that objects with different weights accelerate at the same rate (regardless of gravity). He noted that a ball rolled down an inclined plane would progress up a second plane to approximately the same original height, and would reach the same height were it not for friction. This established an important principle, since much followed in the design of experiments – that a complex system can be broken down into simple components obeying idealised rules. As there will always be friction in the real world, Galileo designed his experiments as if it did not exist, then factored in the effects of it afterwards.

Galileo also made the slope of the second inclined plane shallower and shallower, and came to the conclusion that were it completely flat, and in the absence of friction, the ball would keep rolling forever. In this way it can be seen that just as Kepler had mapped out the route to Newton's discovery of gravity, so Galileo was doing the groundwork for his laws of motion. He understood the idea of a force, as was later incorporated into Newton's laws of motion, and can claim to be the founder of modern mechanics.

As noted, the first telescope was in fact invented by Leonard Digges in England, but this discovery was not taken up elsewhere, and the instrument was reinvented by a Dutchman called Hans Lipperschey in 1608. Galileo immediately understood the commercial and military value of this instrument to his masters in Venice. For one thing, there was money to be made simply by being the first to identify which ships were approaching the port. Failing to meet the Dutchman who had arrived in Italy with a telescope, Galileo at once constructed a superior model himself, with a magnification power of ten times. Very quickly setting it to use, he made a number of startling discoveries. These were:

(1) Jupiter has four moons, now known as the Galilean moons. These are now called Io, Ganymede, Europa and Callisto (many more have since been discovered).
(2) Saturn has a strange oval shape, often described as a globe with ears. This is derived from a side-on view of its famous rings.
(3) The Milky Way is made up of an uncountable number of stars.
(4) The Moon has an irregular surface.
(5) Venus exhibits phases, similar to the phases of the Moon. Its light must be reflected from the Sun, and the phases show that it must orbit the Sun – incontrovertible evidence for the Copernican theory.
(6) The constellation of the Pleiades, a group seven stars named after the daughters of Atlas, now became forty stars.

All this was set out in a little booklet called *The Starry Messenger* (*Siderio Nuncius*) in 1610, which turned Galileo into a European celebrity. For anyone with any common sense, these findings were the end of the road for Aristotle and Ptolemy's view of the universe. A planet with four independent moons was not sitting in a fixed crystal sphere. It would have taken a clever epicycle indeed to describe their motions!

Galileo still had a long career ahead of him, and before long he became the first modern scientist to describe sunspots. He also developed an effective compound microscope. Again, he understood the limits of physics. Certain aspects of a physical entity can be described mathematically – its shape, size, number, position and motion. Other aspects cannot – taste, smell, colour and sound. These belong more in the realm of chemistry, a discipline which beyond the time of Galileo was mainly concerned with the manufacture of medicines.

At first, Galileo's findings did not cause a problem with the Catholic church, but before long they began to be seen as an attack on the established order. This was the problem with the new science – it was established without any need of the cultural baggage of religion, and in fact, it did not need religion at all. The Catholic church felt itself threatened. Galileo was warned not defend the Copernican model as early as 1616, but the production of a book called the *Dialogue* (*Diologo*) in 1632 proved too much for the Papacy. Galileo was tried by the Inquisition, and forced to recant under threat of torture. He was not made of the stuff of martyrs, and because he recanted, he suffered only house arrest. However his recantation was only lip service: at the end of his trial he was said to have muttered *Eppure si muove* (and yet it – the Earth – moves). (Some joke that Galileo was just a victim of jealous peer reviewers!)

With a few notable exceptions, from this time onwards Italy, birthplace of the great flowering of the Renaissance, became a scientific backwater. Its reputation for persecuting scientists – first Bruno, then Galileo – was never really lost. As recently as 2012 geologists were sent to jail there for failing to predict an earthquake.

Late in his life, Galileo made the acquaintance of another Italian scientist, Evangelista Torricelli (1608-47), introducing him to the problem whereby it was found impossible to raise water from a well by more than 30 feet (9 metres) using a suction pump. In 1642 Torricelli conducted his own simulation of the problem using a J-shaped tube filled with mercury, sealed at the top and with the short end immersed in a mercury bath. A similar problem appeared; the mercury settled down short of the top of the tube. This level however varied from day to day, which Torricelli correctly surmised was due to changes in atmospheric pressure. He had invented the barometer – and he had also created a real vacuum at the top of the tube. He was well aware of the significance of the vacuum, something which other scientists, including Descartes, said could not exist.

The French mathematician Blaise Pascal, intrigued by Torricelli's news, constructed his own barometer and sent his brother with it on a climb to the top of the Puy de Dome, almost 1500 meters (one mile) above sea level in central France. Here he found that the height of the column of mercury was markedly less than at normal ground level, demonstrating that the atmosphere of the Earth was likely to run out completely a finite distance from the ground.

Working in quite another field at this time was Jan van Helmont (1579-1644), who lived in what is now Belgium, and a man famous for carrying out an experiment which lasted for five years – he grew a willow tree. First, he dried and weighed the earth in which it was to be planted. After five years he found that the tree had gained in weight by 164 pounds (74 kg), but the soil weighed only two ounces (56 grams) less. He concluded that the tree was made almost entirely out of water. A later experiment by the same man cast doubt on this idea. He burned 62 pounds (27.5 kg) of charcoal, at the end of which he was left with one pound of ash. The fire gave off a gas which was invisible like air, but which would not support a flame when collected in a jar. Helmont concluded that the charcoal must have contained 61 pounds of this gas, which he called spirit of sylvester, and which we call carbon dioxide.

Helmont was feeling his way towards what we would call the carbon cycle, but failed to realise the essential part played by the air itself. Trees take in carbon dioxide from the air and by the process of photosynthesis, extract the carbon and expel the oxygen. The carbon is then used to provide energy for the tree, combining with water to synthesize sugars, and to build its structure. The other inputs to the structure, drawn from the soil, are indeed water, and trace elements. When the wood from the tree is reduced to charcoal, the water is driven off, leaving mostly carbon behind. When this is burned in the air, it combines with oxygen to make carbon dioxide. So the tree is made neither of water, nor of carbon dioxide, but of hydrocarbons – complex organic chemicals composed principally of carbon, hydrogen and oxygen.

Van Helmond went on to distinguish a number of gases – in fact he invented the word. He called a gas "chaos", to him an unordered form of pre-matter. In his thick Flemish accent this sounded to English ears like "gas". Some were combustible, some had pungent odours, others were absorbed by liquids. We can guess what some of these were, but this knowledge did not re-emerge in a codified form until the time of

Priestley and Lavoisier in the 1770s. In fact Van Helmond really gave shape to the very concept of a gas, other than the common air itself, which was not recognised for what it is – a mixture of different gases.

Another contemporary of Galileo was William Harvey (1578-1657), an English doctor born in Folkestone who became the physician to King James I and his son Charles I. Medicine had not stood still during the Renaissance, and indeed the medical part of that movement was the revival of the works of Galen, who had been the physician to the Roman Emperor Marcus Aurelius. Galen was in turn a follower of the original doctor Hippocrates, after whom the Hippocratic oath, still taken by doctors today, is named. Central to the medicine of both was the idea that both the temperament and the health depended on four body fluids: blood, yellow bile (choler), black bile (melancholy) and phlegm. Terms such as phlegmatic and sanguine, used to describe personalities, derive from this.

However, inquisitive surgeons soon found out that Galen was hardly the last word on the subject. One of these, a man from Brussels named Andreas Vesalius, published a number of detailed drawings of human organs, notably in a book called *De Humani Corporis Fabrica* (*On the Fabric of the Human Body*, usually known as the *Fabrica*), published in 1543. After him came Gabriele Fallopio, who described the fallopian tubes (so named after him). In fact he described this link between the uterus and the ovaries as a *tuba* – a trumpet, based on its shape; which someone, somewhere, thought meant "tube"!

Harvey himself famously became the first man to describe the circulation of the blood within animal and so human bodies, published in a book called *De Motu Cordis et Sanguinis in Animalibus* (*On the Motion of the Heart and Blood in Animals*) in 1628. Before Harvey the received wisdom, going all the way back to Galen, was that blood is manufactured in the liver and then is carried in the veins to provide nourishment to the tissues, getting used up in this process, so that new blood has constantly to be manufactured within the liver. Harvey easily disproved this theory by measuring the pumping capacity of the heart – in fact about 260 litres (455 pints) an hour, three times the weight of the average man. Harvey knew that the heart must be a pump, pushing blood at high pressure into the arteries. Blood at this stage is bright red because it has picked up oxygen from the lungs. On its return journey to the heart, it is a much clearer, darker red, its oxygen having been used up to fire the microscopic power stations (the mitochondria) of the cells. It is pumped back along the veins, as distinct from the arteries, which operate valves to make this a one-way movement.

What Harvey could not describe, because microscopes of sufficient power had not yet been invented, was the mechanism of transfer between the lungs, arteries and veins. This missing link was finally supplied by the Italian Marcello Malpighi in 1661. He showed that the lungs, arteries and veins are connected by tiny capillaries or tubules, only visible under the new microscopes.

That the red coloration of blood is due to the presence of iron was suspected by a later Italian, one Vincenzo Menghini of Bologna. In 1745 he roasted 140 grams (five ounces) of the blood of a dog. He was left with nearly an ounce of solid remains. Poking about in this residue with a magnetic blade, he found that most of it was attracted by the blade. So blood not only contains iron – it contains rather a lot of it!

Between Galileo and the next great scientist, Newton, came the life of the Frenchman Rene Descartes (1596-1650). Descartes made a surprising early move by joining the military in Holland in 1618, the year the Thirty Years War began. However, at that stage he was expecting only barrack duties, sufficiently light to allow him to indulge his lifelong habit of rising only at noon, having spent the morning thinking in bed. He later inherited sufficient funds to allow him to travel in Europe and get by on his frugal lifestyle.

He is principally remembered today for two things, his virtually meaningless statement *"Cogito ergo sum"* (I think therefore I am), known to every schoolboy Latin student, and for his Cartesian coordinates, named after him. He dreamed up this idea whilst lying in bed! They describe the $x$- and $y$- axes on any graph, and in three dimensions ($x, y, z$) are capable of describing the position of any object. This was part of his work on analytical geometry, the extention of algebra into geometry which was to provide the basis for the development of the calculus in the next generation. It had never occurred to anyone before that the relationship between two elements in an algebraic equation, $x$ and $y$, could be plotted on a graph, and the shape and meaning of the curves produced in this way began to occupy minds all over Europe. New questions arose, such as how to measure the area beneath the line on the graph, when that line was curved. Newton later devised the branch of mathematics known as the calculus specifically to deal with this problem.

In cosmology Descartes envisaged that the universe was filled with gigantic but insubstantial celestial whirlpools or vortices, through which objects such as comets and planets made their way. Despite the lack of evidence or mathematical rigour, this approach remained popular in his native France long after Newton had given a much better explanation, and held back the development of real astronomy in that country, according to some, for two centuries.

Descartes spent most of his adult life in the Netherlands, but in 1649 he accepted an invitation from Queen Christina of Sweden to join a circle of intellectuals which she was forming in Stockholm. When he arrived, he found he was expected to offer instruction to the Queen at 5 am every morning. After a lifetime of lie-ins, this proved too much for him, and he failed to survive his first Scandinavian winter, dying in February 1650 at the age of 54.

Descartes rejected the idea of a void or a complete vacuum existing anywhere, but contemporaries were beginning to think differently. One of these was Pierre Gassendi (1592-1655), who revived the idea of the atom. He published his own ideas on atoms in a book in 1649. He believed that atoms move around in a void, and that there is literally nothing in the gaps between them. He also thought that they could join together and form molecules. He turned out to be right on both counts, but anything resembling a proof was to be a long time coming. Gassendi is also the man who first observed the transit of the Sun by Mercury.

A real scientist (rather than an occasional philosopher and military man like Descartes) was meanwhile emerging in Holland. This was Christiaan Huygens (1629-1695), who would be much more prominent today had not his career been overshadowed by Newton. He invented the pendulum clock, patented in 1657, which was a great advance in accurate timekeeping (on land, at least, if not at sea). Also during the 1650s, Huygens (together with his brother) developed telescopes which were vastly superior to any previously known, all done by the careful grinding and polishing of lenses. This enabled him to discover Titan, a moon of Saturn and in fact the largest satellite in the solar system. This created a sensation and made Huygens famous throughout Europe. During the 1670s Huygens developed the balance wheel and spring assembly which gave watches a much greater degree of accuracy than previously; however it was not exactly quartz time, but accurate only within about ten minutes a day.

Following on from Descartes, and this time a real astronomer, was Giovanni Cassini (1625-1712), who hailed from Nice, at that time part of Italy; in fact he was to divide his working life between Italy and France. He was the first to discover four of Saturn's moons, including Iapetus, and is credited with Robert Hooke with the discovery of the Great Red Spot on Jupiter. In 1672 he sent a colleague to French Guiana whilst he stayed in Paris, in order to measure the angle of Mars from two widely separated places simultaneously, and by this means compute the parallax and so its actual distance from the Earth. As the geometric ratios of all the planets to one another were already known, this single absolute measurement allowed the approximate dimensions of the entire known solar system to be calculated for the first time. He also discovered the division of the rings of Saturn, called the Cassini division after him to this day. The names of both Cassini and Huygens are remembered in a modern spacecraft, launched in 1997 specifically to investigate the moons of Saturn. It consists of two elements, the Cassini orbiter and the Huygens probe. This latter landed on Titan in 2005, the only spacecraft ever to land in the outer solar system, and continued to send data for 90 minutes after landing.

An associate of Cassini was the Dane Ole Romer. The moons of Jupiter make regular orbits around the planet, in doing so disappearing from view and reappearing at predictable intervals. However Romer noticed that these intervals were not quite as regular as expected. On the basis of a pattern he had detected from previous observations, Romer predicted that the innermost Galilean moon, Io, due to reappear of 9 November 1679, would in fact appear ten minutes late, and he was proved sensationally right. This is because the distance to the moons varies according to where the measurements are taken in the orbit of the Earth, and the time taken for light to reach the Earth from Io varies significantly from the nearest and furthest approaches to Jupiter along the Earth's orbit. Romer had just measured the speed of light. He came up with a figure of 225,000 kilometres per second, not so far from the modern figure of 300,000 km per second – fast, for sure, in fact as fast as anything can travel in the entire universe, but by no means instant. This tiny discrepancy – ten minutes – was later to prove the key to unlocking the way the whole of space works.

Another scientist from this period was the German Otto von Guericke, a man who gave spectacular displays concerning atmospheric pressure in Magdeburg in 1654 (and at later dates). His equipment consisted of two copper hemispheres, an air pump and two teams of horses. The hemispheres were greased together to form a globe, and the air was pumped out of the globe using an air pump which required two men to operate it. When all the air had been sucked out, the two teams of horse, each of eight animals, could not prise the hemispheres apart. Yet when a valve was opened to readmit the air, they fell apart spontaneously, without any need of horses. The force holding the hemispheres together was simply the difference between the atmospheric pressure on the outside and the vacuum inside – there was no opposing pressure inside the globe. This demonstration was repeated all over Germany and astonished crowds wherever Guericke took it. In fact, air pumps and vacuums were to play an important part in science in the nineteenth century, and many advances in tube technology, including cathode ray tubes and radio valves, can be attributed to the later invention of a better vacuum pump.

Another German from this period is Hennig Brand (1630-92), the man who discovered the first new element since the Middle Ages, phosphorus – and he did this in a very strange way. His fist step was to collect fifty buckets of human urine from a nearby barracks, which looked a promising yellow colour – of course, Brand was looking for gold. He then allowed the urine to evaporate and putrefy until there was nothing left but a pasty residue. This he then heated in a retort with a double measure of sand – also a promising colour – collecting the fumes through water. The final distillate he found under the water was a transparent waxy substance. When removed from the water it glowed into the dark, and proved so volatile that it could ignite spontaneously. Soon a whole industry was set up to make phosphorus, no doubt causing hasty changes to property deeds as this process must have been even worse than that other famous stinker, tanning. Winning phosphorus in this way, however, was so difficult that it was used for little more than party tricks for over a century.

Our next scientist is Robert Boyle (1627-1691), considered to be the founder of modern chemistry. His father, the Earl of Cork, though not a traditional aristocrat, was one of the richest men in the country, having made his pile amidst the constant troubles of Ireland. Hence Boyle Junior was able to pursue the career of gentleman scientist, devising and funding his own experiments. In the course of this work he published a number of books, including *The Spring in the Air* and *The Skeptical Chymist*, which attracted widespread attention. A retiring gentleman, Boyle refused many honours, including the Presidency of the Royal Society and a peerage of his own. His work on gases demonstrated Boyle's law, which states that the pressure and volume of a gas are inversely proportional. This implies that a gas is a flexible entity which can be compressed – it must in fact consist of separate particles moving about in a void. If it is compressed then its volume decreases. The before- and after- pressures and volumes of a gas are linked in the formula:

$$p_1 V_1 = p_2 V_2$$

Therefore any increase from $p_1$ to $p_2$ must be accompanied by a corresponding decrease in $V_1$ to $V_2$. Twenty years later the same law was promulgated by a Frenchman called Mariotte, who added the important provision that the temperature of the gas must remain constant, something certainly realised but not stated by Boyle. Hence Boyle's law is known as Mariotte's law throughout Europe. If a gas is heated, it expands on its own – the principle behind the steam engine.

Picking up on the ideas of the Frenchman Gassendi, Boyle also espoused a form of atomic theory, believing that all matter is made of tiny particles which move around in liquids and gases, but whose position is fixed in solids.

A strange fellow now appears on the scientific stage, Robert Hooke (1635-1703), a man who seemed to have a finger in every pie in the great scientific revolution of the seventeenth century. Born on the Isle of Wight, he studied at Oxford and became assistant to, amongst others, Robert Boyle, for whom he created vacuum pumps for experiments with gases. From 1662 he was the curator of experiments at the newly formed Royal Society of London, a position which gave him great influence. He came close to establishing the inverse square law of gravity, eventually elucidated by Newton, and several other things, but never seemed able to come up with the hard experimental evidence or mathematics to back up his hunches. A cantankerous and scheming individual, he clashed with the overbearing Newton. Today he is mainly remembered for his remarkable book *Micrographia*. He used the newly improved microscopes to produce large-scale drawings of familiar creatures, such as fleas and lice, which astonished the public and kept the diarist John Pepys awake at night wondering just what had bitten him. The book includes the first description and diagram of a cell, so-called by Hooke because a collection of them reminded him of the monks' cells in a monastery.

The next great figure in astronomy was born only shortly after the death of Galileo. This was Isaac Newton (1643-1727), arguably the greatest Englishman who ever lived, and whose advances in science appear almost supernatural, though he himself said that if he had seen further, it was by standing on the shoulders of giants – by which he meant Kepler, Galileo and Descartes. Newton was born at Woolthorpe near Grantham in Lincolnshire, the son of a prosperous farmer who died before Newton was actually born. When he was only three years old, his mother remarried, and the young Isaac was left to a gloomy and lonely upbringing with his grandparents, only re-entering under his mother's roof when her second husband died, when he was eleven years old. He then attended grammar school in Grantham, lodging in lively company with a kindly apothecary, Mr Clark.

When he left school, his mother tried to get him to run the farm, but he made a very poor fist of it, and was fined several times for allowing his animals to graze on the crops of his neighbours. Evidently his heart was not in the job, and he was known frequently to take a book with him into the fields. On Saturdays, sent by his mother into Grantham with a servant to conduct business, he would leave the jobs to the servant and bury himself all day in Mr Clark's books.

However he did have two connections to Cambridge. One was his uncle, William Ayscough, who was a Cambridge graduate, and the other was a brother of the woman with whom he lodged at the grammar school in Grantham, Henry Babbington, a fellow of Trinity College. So it was that Newton's mother reluctantly agreed to allow Isaac to attend university at Trinity College, on a miserable allowance of £10 a year (when her own income was £700 a year). The servants and farmhands in Wooltharpe were pleased to see him go, considering him "only fit for the Varsity".

Newton arrived in Cambridge at the age of 18 in 1661, finding himself two or three years older than the average freshman, and out of his depth socially (an experience still felt today by many provincial students of relatively limited means). He was apparently miserable at first, but by early 1663 he had teamed up with one Nicholas Wickins (from Manchester), with whom he shared rooms for the next twenty years. It is sometimes assumed today that this relationship was homosexual, as Newton rarely showed anything but a horror of women and their sexuality. However, when he finally left Cambridge, Wickins did marry and produce a family.

Newton then prospered at Cambridge, becoming a fellow of Trinity and then Lucasian Professor of Mathematics at the age of 26. However for most of the two years between 1665 and 1667 he was obliged to retire home to Woolthorpe because Cambridge University was closed on account of an epidemic of plague. According to legend, it was here, in his mid-twenties, that he had his "eureka" moment when witnessing an apple fall from a tree. It occurred to him that the apple did not just fall – it was pulled to the Earth by a force, the force of gravity, or weight – the opposite of levity, or lightness. He concluded that the force or influence of gravity also held the planets in place in their orbits around the Sun. This force is governed by an inverse square law – as the distance between heavenly bodies increases, so the force diminishes exponentially. The force of gravity twice as far from its source is spread out over twice the area and hence weakens to a quarter of its strength. Other forces emanating from a single central source also follow an inverse square law, including electric current.

The source of gravity is mass, and it operates at a distance without the need for any medium in which to work; in other words it works in a vacuum as exists in outer space. In the case of the Earth and the Moon, the mass of the Earth acts as the force which maintains the Moon in its orbit, but that force spreads out and weakens in all directions from the Earth itself. Newton eventually concluded that the Moon would move in a straight line (away from the Earth) were not the single force of gravity sufficient to make it move in an orbit.

This is Newton's formula for F, the force of gravity:

$$F = g(m_1 m_2)/r^2$$

where $m_1$ and $m_2$ are the two masses, r is the distance between them and g is the "gravitational" constant. This certainly goes along with the principle that correct theories are characterized by both simplicity and elegance, and in that respect it is similar to Einstein's relativity equation.

Newton found himself able to calculate the orbits of the planets on a purely mathematical basis. To assist him in his calculations, he actually invented his very own new mathematics, today known as the calculus (he called it fluxions). This enabled him to calculate the area under a curve, or inside an orbit. There eventually arose a priority dispute with the German mathematician Gottfried von Leibnitz, who invented virtually the same system independently and apparently shortly afterwards. Newton was an abrasive man to take on, and Leibnitz found himself worsted by the Newton steamroller, though it is his form of notation for the calculus which is used today.

As part of this work, Newton formulated his three laws of motion. These state:

(1) That if a body experiences no force, then its velocity remains constant, either zero if the body is stationary, or in a straight line and at constant speed if the body is moving. This law is difficult to see in operation on the Earth where friction means that moving bodies always slow down, but it applies much more directly to heavenly bodies which are moving through empty space. This law derives directly from the work of Galileo and his inclined planes. It is one of the fundamental laws of classical mechanics, also known as the principle of inertia.

(2) The acceleration of a body is parallel and directly proportional to a force acting on that body, and it is inversely proportional to the mass of the body. That is, if a billiard ball is stationary on a billiard table, and it is hit dead-centre by another billiard ball, it will move straight forward, but it will not move as fast as the ball which hit it, because its own mass lessens the

impact. If the stationary billiard ball were twice the weight of the ball which hit it, it would only move half as far.

(3) When a body exerts a force on another body, then the second body also exerts a force which is equal and opposite on the first body, or to put it another way, to every action there is an equal and opposite reaction. Thus when one moving billiard ball hits another stationary ball head-on, the second law means that the second ball moves away, but the third law means that the first ball is slowed down or stopped by the reaction of hitting the second ball.

Newton's documented his findings in natural philosophy (today called physics) in the most comprehensive manner in his publication of 1687, *Philosophiae Naturalis Principia Mathematica*, often known as simply Newton's *Principia*. Hardly a work of popular science, Newton deliberately set the book above the level of the common reader in order "to avoid being baited by little smatterers in mathematics." He was persuaded to publish the book, twenty years after the basic research in it had been completed, by Edmund Halley, later to become the Astronomer Royal, who helped to fund the considerable publication costs from his own pocket. The Royal Society, which should have been able to fund the publication, had used up its entire budget on a costly flop, John Ray's *History of the Fishes*, of which more later. (Halley himself had just accepted a position as clerk to the Royal Society, which could no longer afford to pay him the promised salary of £50 per annum. Instead it gave him some spare copies of *History of the Fishes*.)

The *Principia* also included two conjectures which were to provide some interesting job opportunities in future years. One was that the Earth is not a perfect sphere. The centrifugal force of its spin will cause a slight flattening at the poles, and a bulge at the Equator. This in turn means that the length of a degree of longitude, or meridian, is not exactly the same over the whole globe – it increases towards the poles. Much effort was later to be expended, notably by the French, in trying to obtain exact measurements of a degree of longitude. Secondly, Newton said that a plumb line suspended near a large mountain would be deflected towards it by the gravitational mass of the mountain. In fact the movement of the plumb line from the vertical could be used to measure the gravitational constant, and so the mass of the Earth. Nearly a century later (1774) the Astronomer Royal, Neville Maskelyne, made the appropriate measurements for a mountain in Scotland, Schiehallion, which is very regular in its pyramidical shape. It eventually fell to the

English academic John Mitchell to devise an easier way of doing this same job using lead balls (of which more later).

Newton also made great advances in the field of optics, devising his own experiments using glass prisms, said to have been bought for a penny at Stourbridge Fair. He demonstrated that white light is made up of a spectrum of colours which can be seen by focusing the white light through a prism. He was hardly the first person to note the light split up (or refracted) by a prism in the this way, but his predecessors had thought that the prism itself produced the colours. Newton said no – the colours already exist in the light, and are merely separated out by the effect of the prism. To demonstrate this he isolated one colour of light by passing a beam through a prism and then a slit in a card. This light – say, just the green light – he then passed through a second prism, the output from which was still just green. He had in fact excluded all other wavelengths.

He also showed that the light split up in this way can be recomposed into white light by focusing it through a lens and a second prism. In his theory of colour, he deduced that colour comes from light, rather than being an intrinsic property of any object, which is given its colour when light shines upon it. Even an orange does not look orange in a dark room! Newton took a "corpuscular" view of light, regarding it as a stream of weightless particles emanating from the Sun. There was later to be much debate about whether light acts in this way, or as a wave.

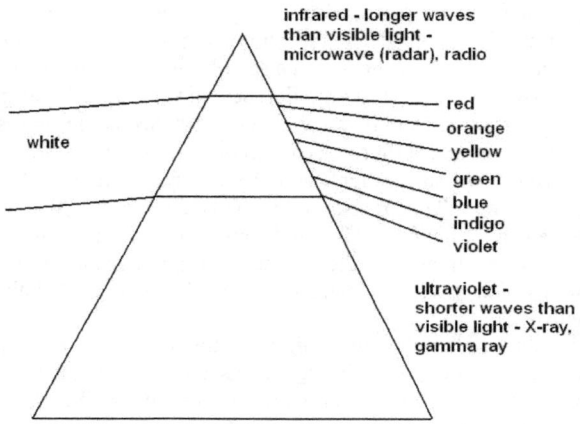

White light as split by a prism

In his studies of light and in order to observe the universe, Newton also made the first successful reflecting telescope. Others by this time had started to address the problems with telescopes. Refracting telescopes, which bend the light, suffer from "chromatic aberration" as the light is dispersed into the different colours of the spectrum. Newton eventually wrote up his many investigations into light in the book *Opticks* (1704).

Newton was briefly a member of Parliament in 1689-1690 and 1701. In 1696, at the age of 54, he moved to London to take up the position of warden of the Royal Mint, which he occupied with enthusiasm. In fact he did little real work in science from the age of 35, spending most of his time on hopeless alchemical experiments (attempting to find gold), Bible studies and the like. In religion he was (like Bruno) an unorthodox Arian, though still deeply religious for all that. He was also a very disputatious man, and got into a number of well-known spats, notably with Leibnitz and also with Robert Hooke. In later years his social life was much softened when he shared his house with the daughter of his step-sister, Catherine Barton, who acted as his hostess in social matters. This young lady was eventually to inherit a fortune from the politician the Earl of Halifax, friend and sponsor of Newton and one of the founders of the Bank of England (when he was known as Charles Montague). Tongues wagged.

Even in his own day, Newton was recognised across the civilized world as a genius, an opinion not altered one whit in subsequent times. The poet Alexander Pope composed an epitaph:

> Nature and Nature's laws lay hid in night;
> God said "Let Newton be", and all was light

Newton's rival in the invention of the calculus, Gottfried von Leibniz, was equally generous in his praise, stating that Newton's contribution to mathematics was equal to all the accumulated mathematics which had gone before.

## Chapter 4 – Into the Eighteenth Century

In this great age of science, which saw the opening of the Royal Society in 1662, the Royal Observatory at Greenwich was also opened in 1676. Along with it was created the new post of Astronomer Royal, held by John Flamsteed (1646-1719) from the opening of the observatory until his death. He was a very jealous man, delaying the publication of the astronomical tables he was employed to produce, sometimes for years, at every opportunity. For many years his great rival was Edmund Halley (1656-1742), who had a long and active life as a mathematician and astronomer, and who was closely linked with Newton. The issues between them were not resolved until Halley took over himself as Astronomer Royal after the death of Flamsteed. However, by that time Halley had compared accurate modern measurements with astronomical maps made by the Greeks. These showed that although most stars remained in fixed positions, some had moved, notably Arcturus, a bright star in the constellation Böotes whose position in the sky had somehow shifted by a whole degree (in fact, it is one of the closest stars to the Earth at a distance of 36 light years, making any relative movement much more likely to be detectable).

Earlier in his career, Halley had at one time set out to investigate the possible causes of the Biblical Flood of Noah, thought at that time to be responsible for the deposition of all sedimentary rocks. In the thinking of the day, the creation of the Earth was fixed at a date of 4004 BC, as calculated by one Bishop Ussher in 1620 by counting back the generations to Abraham and eventually to Adam and Eve in the Bible. By comparison with the processes of erosion taking place in his own times, Halley came to the conclusion that the Earth must be a great deal older than that (the age of the Earth is now thought to be 4.55 billion years). He was right, of course, but his beliefs gave him a name as a heretic. Although he was unlikely to suffer the fate of Bruno, such unorthodoxy was a barrier to official appointments in the universities.

Halley also began to gain some appreciation of the properties and size of atoms. He started by comparing a lump of gold with a lump of glass, finding that though they were of the same volume, the gold weighed about seven times as much as the glass. He came to the conclusion that the glass must contain much more void space than the gold, perhaps because its atoms were less tightly packed. He also experimented with gold leaf, which he thought was probably just one atom deep, and concluded that there must be millions of atoms in even the tiniest piece of visible gold. He calculations were in fact well short of the true number, but they were on the right lines.

Halley used Newton's new laws to calculate the orbit of the comet which had appeared in the skies in 1682, predicting that, with a periodicity of 75-6 years, it would reappear in 1758; it did (just, on 25 December), when it was named after him. This was one of the first manifestations which could actually be used to prove Newton's laws, still questioned in some countries. An eclipse of the Sun was used to similar effect in 1919 after Einstein had promulgated his General Theory of Relativity in 1915. Halley also recognised that the transit of Venus across the face of the Sun could be used to measure the distance between the Earth and the Sun, if this was measured from distant points on the face of the Earth (to obtain parallax measurements as a basis for triangulation). He calculated that the next transits would take place in 1761 and 1769. In fact when the time came, observations were taken from more than sixty points around the world, one of them by Captain Cook on Tahiti. This gave a distance to the Sun of 153 million kilometers/95.6 million miles (the modern measurement is 149.6 million kilometers/93.5 million miles). In this way Halley made his last great contribution to science 27 years after his death.

In this era there were also developments in botany and zoology. The first name of note is the English physician Edward Tyson (1650-1705), generally regarded as the founder of comparative anatomy, which is the study of the physical relationships between different species. One of his most memorable dissections concerned a porpoise which had swum up the Thames and found its way onto a fisherman's slab, where a curious Tyson bought it for the not inconsiderable sum of seven shillings and six pence (later reimbursed by the Royal Society). He found to his amazement that this "fish" was in fact a mammal, with an internal structure and skeleton like any land-based quadruped. He found, for example, that the fore-fin had a structure with a spatula (shoulder blade), humerus (upper arm), ulna and radius (lower arm), carpus (wrist), metacarpals and five digits, and announced his findings to an astonished Royal Society. Tyson also dissected a chimpanzee (which he called an orang-utang) and found it more similar to a human than a monkey.

A genuine biologist (rather than a medical man) at this time was John Ray (1627-1705), who published a number of multi-volume illustrated books on the history of fishes, plants and insects. He worked in collaboration with his sponsor and friend Francis Willughby (1635-72). From a poor family in Essex, Ray made his way to Cambridge where he became a fellow of Trinity College. A mild but determined Puritan, he refused to swear an Anglican oath under the Act of Uniformity of 1662, thereby losing his post at the university and condemning himself to a life dependent on charity. After the execution of King Charles I and the rule of the English Puritans under Oliver Cromwell in the period 1649-60, the monarchy had been restored under King Charles II in 1660. This brought about a strong and nationwide reaction against Puritanism and in favour of the Anglican religion, and Ray was caught out by this; but in fact most of his colleagues paid lip service to the new oath, and only a dozen from Cambridge besides Ray chose to defy the oath. Ray was in fact just one in a long line of scientists from this earlier period whose views on religion got them into trouble with the authorities.

The charity now need by Ray was supplied willingly enough in his shorter lifetime by Willughby, after whose death Ray had to scratch a living somehow whilst compiling his massive books. It was his *History of Fishes*, credited to Willughby, which soaked up all the funds for publications at the Royal Society, meaning that nothing was available from the Society for Newton's *Principia Mathematica*. Of greater later importance was his *History of Plants*. He discovered certain fundamental principles, such as that there are two types of seed, the dicot (such as a bean) which split into two, and the monocot (such as maize or corn), which form one single mass. He also realised that there could be no simple way of classifying plants – that would depend on taking all their characteristics into account. His successor, Linnaeus, attempted a simple scheme by using only the sexual characteristics of plants, but his own successors had to go back to Ray. Considered by modern botanists to be a true genius whose ideas turned out to be right, it would take several generations for this to be recognised. One problem was the limited funds available for his publications – *History of Plants* was not even illustrated for this reason; they did not sell well.

At a certain stage, every science requires a genius of classification, but in biology, this was a two-stage process. Ray turned biology into a systematic study of plants and animals, bringing order and logic to the classification of species and providing the documentation required by the next great classifier, the Swede Carl Linnaeus (1707-1778).

Initially pursuing a career in medicine, this man soon found himself involved in botany, where he became very curious about the relatively new idea that plants reproduce sexually, and have male and female sexual organs (stamen and pistil) which don't look a bit like the animal equivalents. He came to use the differences between the reproductive parts of flowering plants as one of the ways of cataloguing and classifying these plants. It is for this line of work that he is still famous – taxonomy, the naming and classification of plants and animals. His *Systema Naturae*, first appearing in 1732, ran to many editions. This work introduced the binomial system of nomenclature, so that every species has a two-word name, for genus and species. Here he also defined new terms, including *Mammalia, Primates* and *Homo sapiens*. Altogether Linnaeus provided names, descriptions and a place in his classification for 7700 species of plants and 4400 species of animals. His work caught on elsewhere in Europe because he consistently classified by physical characteristics only. Before his time there were many rival systems – Buffon, for example, classified his animals by their utility to mankind. Also, Linnaeus was simply good at taxonomy, having the happy knack of selecting the critical qualities which bring species together, or separate them from other species. He was only too well aware of his own abilities, describing himself as the "prince of botanists".

Somewhat controversially, though in line with the spirit of the times, he provided man with his own genus, *Homo*, when on any more objective basis (especially given the high level of shared DNA as nowadays determined) he should really have lumped us in with the chimpanzee genus *Pan*.

The writer Natalie Angier suggests a mnemonic line to remember the full Linnaean order of classification: "Kings pour coffee on fairy god-sisters": kingdom, phylum, class, order, family, genus and species. (In fact the term *phylum* was not used until 1876.) In the case of people, the genus and species is *Homo sapiens*; then joined by the apes in the family Hominidae. These hominid apes then meet the monkeys, lorises, lemurs and the like in the order Primates, and in the next level up another 4,600 or so creatures classed as Mammalia, the mammals, anything from a duck-billed platypus to an elephant. This group is then part of the phylum Chordata and the subphylum Vertebrata, creatures with a backbone, including reptiles, amphibians, birds and fish. Beyond that is the kingdom Animalia. Other phyla include the molluscs and the arthropods (which include shrimps and insects).

Linnaeus also got involved in the increasingly loud questions been raised about the age of the Earth, as more and more evidence came to light about fossils, such as apparently extinct sea creatures found on the tops of mountains. One early pioneer in this field was the Dane Niels Steno (1638-1686), who felt from the evidence provided by modern and fossil shark's teeth that the Earth must have been inundated many times. The idea of a Biblical Flood, lasting no more than 200 days, was beginning to look tenuous, and some scholars were prepared to admit that there must have been several floods. Also. at this time, knowledge of China was increasing, and it did appear that the Chinese civilisation stretched back beyond the beginning of Biblical time, that is, 4004 BC.

A contemporary of Linnaeus was the quaintly-named Frenchman, the Compte de Buffon (1707-1788), a wealthy man able to devote his early years to touring Europe in the study of nature. Things were not entirely straightforward for him, however, as when his mother died he had to take his own father (who had remarried) to court for his own direct share in her inheritance, a contest he won at the age of 25. From 1741 onwards he was the superintendent of the Jardin du Roi, the King's botanical garden in Paris. All this led him to produce a monumental work of natural history, his *Histoire Naturelle*, which ran to 44 volumes! This did not mark any great leap forward in knowledge, but it did bring together a vast amount of material, and gave it coherent shape. Buffon was another to comment upon that increasingly vexed subject, the age of the Earth. He thought that the Earth had probably formed from molten material thrown out of the Sun, and if so could not be less than 75,000 years old.

Buffon was one of the first people to state publicly, in his *Histoire Naturelle*, that the idea of Creation as set out in the Bible must be wrong. According the fundamentalists – and in these times, that meant most people – all types of creatures and plants were created at once by God, as described in the book of Genesis, and had not changed since. Bouffon stated what seemed obvious, that the horse and the ass had sprung from a common ancestor, as had man and the apes, and the same applied to all families of plants and animals. He also noted that some creatures are NOT perfectly designed by an intelligent Creator. This idea, going back to Aristotle, was that every feature of a plant or animal had a purpose, from a cat's whisker to a rose's thorn. So why, for example, did a pig need toes, the bones of which are perfectly formed, but which are of no use to it as they are fused into hooves? Equally, anyone setting out to design a bat would not give it the skeleton of the bats we know, especially the wings, which were clearly "designed" for purposes other than flying. The fact is that any organism carries with it the baggage of its history, and will make the best of what it has got to meet new circumstances.

Another prominent Frenchman from the eighteenth century was the mathematician Jean Fourier (1768-1830). His work sought to describe common phenomena mathematically, so that for example he was able to decompose a sound into a collection of simple sine waves. He also worked out equations to describe the way heat flows from a hot object to a cooler one. This led him to make his own estimate of the age of the Earth, taking into account that early in its history, its crust must have formed, slowing down the internal cooling process. He wrote down his equation for this in 1820, but did not trouble (or dare) to publish the answer, but we can see what this was by filling in the appropriate heat flow numbers: 100 million years.

The eastern part of modern France, known as Franche Comte, was not actually part of France in the eighteenth century, but fell under the dominions of the Duke of Wurttemberg in Germany. It was here in the town of Montbeliard that Georges Cuvier (1769-1832) was born, in the year 1769, which also saw the birth of both Napoleon and the Duke of Wellington. Nevertheless the young man made his way into France, where he became associated with an aristocratic family. Having somewhat precariously survived the French Revolution which began in 1789, Cuvier became a comparative anatomist, and can claim to be one of the first paleontologists, or fossil specialists. His work was at the Museum of Natural History in Paris.

Cuvier saw that predators and prey are built on completely different lines. A meat eater needed fast legs, sharp teeth and claws. A plant eater needed flat teeth and hooves. If an animal had hooves, it could not be a predator; hooves were needed for a fast getaway. No doubt with some Gallic exaggeration, he claimed that an expert (like him) could reconstruct an animal given a single bone from its body.

He also saw no link in body plans between different groups of animals, and so derived four fundamental families: vertebrates, molluscs, radiates (radially symmetrical sea creature such as echinoderms and jellyfish) and articulates (insects and crustaceans). This classification has now been completely superseded, but one cannot blame Cuvier for claiming that there could not possibly be any link between a mollusc and a vertebrate. In fact there must have been at some stage, but it is certainly invisible in the geological record, as both types are present from the Cambrian explosion of marine life (542 million years ago) onwards.

The geological record by this period showed two things clearly – first, that there was a time before there was any visible life, and second, that some forms of life had become extinct. Still Cuvier clung to the view that all life forms had existed from the outset, but some had since died out. He did not accept evolution, or the change from one life form to another across the generations.

## Chapter 5 – The Emergence of Chemistry

After Robert Boyle, many years elapsed before there were any advances in chemistry. One problem was that chemists had no reliable, controllable source of heat – they couldn't just light a Bunsen burner; and controlling heat to obtain measurable results from chemical reactions was one of the keys to progress. However, at least they had a thermometer! This mercury version of this (which is still in use) was invented in 1714 by Gabriel Fahrenheit (1686-1736), a man from a German family who lived in Holland.

Cobalt had been known indirectly for centuries because of the blue coloration of a preparation of its oxide known as smalt. This had a long usage in stained glass and in the famous Ming porcelain from China and subsequent European derivatives from Delft, Meissen and the English Potteries. The basic ore, smaltite, was mined in the mountains between Bohemia and Saxony which have historically been the source of so many minerals in Europe. The miners disliked it intensely because of the harmful arsenic fumes given off when the ore was roasted. They laid the blame on a little earth demon called Kobold. In 1735 a Swedish chemist called Georg Brandt isolated cobalt from smalt, and gave it a name based on the old miner's devil.

In 1751 a new element was discovered by the Swede Axel Cronstedt. Miners from Germany had long been familiar with an apparently useless ore, like copper ore, but which tinted glass green instead of blue when dissolved in acid. They called it Kupfernickel – "Old Nick's copper", as it were copper bewitched by the devil. Cronstedt isolated this substance, which he found had properties quite unlike copper – it was hard, silver-white and attracted by a magnet. He called it nickel, a substance which was eventually to find its place in the manufacture of, amongst other things, coins and stainless steel. So the American five-cent coin, the nickel, is in fact named after the devil. Cronstedt was also a pioneer in the use of the blowpipe, capable of focusing a strong flame which caused the materials subjected to it to give off a characteristic colour. This proved a very useful diagnostic aid in identifying unknown lumps of ore.

The next name in chemistry is the Scot Joseph Black (1728-1799), a scientific contributor to the Scottish Enlightenment of the eighteenth century, which also produced such figures as Adam Smith (economist), David Hume (philosopher) and James Hutton (geologist). Black was the man who first identified and isolated the gas carbon dioxide, and he did this by burning limestone, which is not caustic, to produced quicklime, which is:

$$CaCO_3 = CaO + CO_2$$

Carbon dioxide (which Black called fixed air) is a by-product of this reaction. Black found that it is denser than air, and will support neither flame nor animal life. The next reaction in the chain is the addition of water to produce slaked lime, or calcium hydroxide:

$$CaO + H_2O = Ca(OH)_2$$

Black carefully weighed all his materials before and after his experiments, a technique later taken up with successful results by Antoine Lavoisier.

Black also ventured into physics, investigating something which was of increasing interest in the days of the first steam engines – the nature of heat. He discovered latent heat, the amount of heat required to melt a solid into a liquid at the same temperature, or to evaporate a liquid into a gas, thereby demonstrating that there is an important difference between heat (a great deal of which is required) and temperature (which does not change). In fact it takes the same amount of heat to melt ice into water at 32 degrees Fahrenheit (0 C) as it does to raise that same volume of melted water at 32 degrees Fahrenheit to 140 degrees Fahrenheit (60 C). So it is that melting solids and evaporating liquids must extract heat from their surroundings to change mode. The reverse process takes place when, for example, water vapour condenses back into water droplets – its latent heat is released into the atmosphere. This is the mechanism which gives hurricanes their power – their enormous energy comes from the latent heat of newly condensed water vapour.

Black also defined specific heat, a measure of heat capacity and nowadays defined as the amount of heat required to raise one gram of a substance by one degree centigrade. Black found that different substances have their own specific heat value, so that, for example, it requires a lot more heat to raise the temperature of water by one degree than is the case for iron.

One of Black's university technicians was a young man by the name of James Watt (1736-1819). The first practical steam engine, designed to pump water out of mines, had been invented by Thomas Newcomen (1664-1729) in 1712, and although it was all there was for sixty or more years, it is noted as much for what was wrong with it as for what worked. In principle it has a vertical cylinder with a piston inside it, which is in turn attached to a beam. At the other end of the beam is a counterweight, so the default position is for the piston to be up, and the counterweight down. Newcomen's design worked by forcing steam into the cylinder, then cooling it with cold water to create a partial vacuum beneath the piston which sucked it down. Newcomen engines were widely installed and were of immense technological importance as they were the forerunners of much that was to come. A locomotive is basically a steam engine on wheels, so something which was never meant to move was to develop into the world's first mass transport mechanism, an outcome that Newcomen could hardly have predicted.

Whilst working as a technician at Glasgow University, Watt was given a model of this engine, but found that after only a few strokes it ran out of steam. This was partly because it was a scale model – small objects have a larger relative surface area across which heat can escape. So Watt improved the design by adding a second cylinder, because he could see that a lot of steam and energy was wasted in heating and cooling the single cylinder. In this design there are hot and cold cylinders, each with its own piston, connected by valves, and working in such a way that the hot cylinder stays hot, and the cool one cool. This second cylinder is known as the condenser. Watt later added other refinements including an automatic governor to shut off the steam if the engine ran too fast.

By developing his steam engine and the patents for it in association with a university, Watt was pioneering a modern method of research, but the fact is that he made no commercial progress in this way. According to a contemporary, Dr Johnson, "The noblest prospect a Scotchman ever sees is the high road to England," and Watt took that road. He moved to Birmingham in 1774, where he became part of a scientifically-minded group of people who called themselves the Lunar Society (as they met once a month). This group included such luminaries as Joseph Priestley, Erasmus Darwin and Josiah Wedgwood (these last two the grandfathers of Charles Darwin). Watts' new design became a runaway success after he teamed up with the manufacturer

Matthew Boulton. Watt himself became so famous that the SI (International System) unit of electric power is named after him, and his name crops up every time we need a 100-watt light bulb.

It was a member of the Lunar Society who made the next great leap forward in chemistry, Joseph Priestley (1733-1804), a man curiously not as famous in England today as he is in the United States. His chief claim to fame is that he discovered oxygen, but he also discovered another ten gases, including ammonia, hydrogen chloride, nitrous oxide (laughing gas) and sulphur dioxide. Working on a tip from Henry Cavendish, Priestley collected his gases over mercury instead of water, in which they dissolved. However, he was a much better experimenter than he was a theorist, and having discovered oxygen in 1774, he did not fully understand what he had found; so he had to go to Paris to ask Lavoisier, who understood immediately.

Priestley was a controversial man in his time. From the Leeds area of Yorkshire, he trained as a nonconformist – Calvinist – minister. Even though he abandoned the traditional Holy Trinity early in his career, to become yet another Arian in the manner of Bruno and Newton, he continued in his career in the ministry of believers who took a similar view – the Unitarians – all the while conducting his experiments. For a time he came under the wing of a rich patron, the Whig politician Lord Shelburne (briefly Prime Minister in 1782), who indulged his scientific work. He also married into a family made wealthy by the Industrial Revolution. His wife was Mary Wilkinson (married 1762), the sister of John Wilkinson, the Birmingham ironmaster who made a fortune from the manufacture of arms, notably cannons. However it was Priestley's flirtation with the French Revolution which led to a mob attack on him and his associates in Birmingham in 1791, causing Priestley finally to emigrate with his family to the United States in 1794. By this time sixty-one years old and a very famous man, he was feted on his arrival in New York, but elected to live as quietly as possible in Pennsylvania.

*An Experiment on a Bird in an Air Pump* by Thomas Wright of Derby, a famous painting of the era of the Lunar Society. From the Tate Gallery, London.

It is a curious fact that Priestley was assisted in his work by living near breweries, which produced vast amounts of carbon dioxide as a by-product of the fermentation process. Priestley even managed to bottle this by dissolving it in water, thereby inventing soda water. It soon caught on all over Europe, but Priestley refused to patent his discovery. It was taken up by one Joseph Schweppe, a Swiss immigrant, who established a soda water business in London in 1792, still with us today of course.

When Priestley discovered oxygen, he and his contemporaries were very much confused by the "phlogiston" theory of the times. This held that when a metal is burned in the air, it gives off phlogiston to leave a metallic ash called a calx behind. Priestley heated pure mercury in the air, leaving a calx of what we now know is mercuric oxide behind. He then reversed this process by focusing strong sunlight through a lens onto the calx, whereby it reverted to metallic mercury, at the same time releasing a new "air" – in fact oxygen. Experiments with the new air showed that, unlike carbon dioxide, it is very good at sustaining life (mice lived much longer in a limited supply of this than they did in the same amount of ordinary air), and is also very inflammable. Priestley thought of this substance as "dephlogisticated air" – air with the supposed product of the combustion of the mercuric calx taken out of it, as the mercury had resumed its fully phlogistic form after this gas had been given off. He also thought that because objects burned so readily in his new air, it must be air which was devoid of phlogiston, which was the reason it absorbed it so readily.

Working independently in Sweden in the 1770s, the pharmacist Carl Scheele also discovered oxygen. Priestley and Scheele used the same experiment with mercuric oxide, as this substance readily gives up its oxygen and reverts to mercury. However it is Priestley to whose name is attributed the discovery of the gas, as he passed it into the scientific and eventually public domain by giving details to Lavoisier. Scheele is thought to have discovered a number of new elements and compounds, but he did not always pass his findings into the public domain. Also he became the first person to describe chlorine gas, but he did not identify it as a new element. This was left to Humphry Davy, 36 years later. In the meantime chlorine had already been developed in France as the first effective bleach.

One thing that Scheele did pass into public knowledge was his discovery, with another Swedish scientist, that phosphorus is a major constituent of bones. It was this source of supply (rather than the previous one, urine) which made more phosphorus available, which could then be used in applications such as phosphorus lamps. However, Scheele had a habit of tasting, inhaling or otherwise ingesting every new chemical he came across. He was found dead at his desk at the age of only 43 in 1786, surrounded, according to legend, by bottles of deadly chemicals such as hydrocyanic acid and under a cloud of chlorine gas!

A brand new metal was discovered in the 1780s in the Austrian empire in part of what is now Romania. This was tellurium, found in association with gold ores. This metal was to become notorious amongst chemists because it so readily forms a compound with hydrogen. This has the same type of bad eggs smell as hydrogen sulphide but is ten time worse and lingers for days!

Active at the same time as Priestley was Henry Cavendish (1731-1810), a retiring gentleman scientist from one of the richest families in the country – his grandfathers were the Dukes of Devonshire and Kent. An intensely shy and silent man, he left written instructions to his housekeeper and avoided all women. When called upon by a director of the Bank of England, where his deposit was the largest in the entire bank, he sent him away with a sharp instruction to get on with his job and never to bother him again. For the sake of his science, he was prepared to make forays as far as the Royal Society, where he was observed never to say a word, but to utter a shrill little cry as he shuffled from room to room.

Many of his experiments involved sulphuric acid ($H_2SO_4$), a corrosive chemical which was known to the alchemists as oil of vitriol. It can be formed by burning raw sulphur, which exists in nature. From the sixteenth century it was prepared in bulk by burning sulphur with saltpeter (potassium nitrate, $KNO_3$) in the presence of steam. A typical school experiment uses the acid as follows:

$$CaCO_3 + H_2SO_4 = CaSO_4 + H_2O + CO_2$$

Calcium carbonate (in the form of chalk or limestone) plus sulphuric acid gives calcium sulphate (a salt), water and carbon dioxide. Cavendish carried out this experiment, in his terms combining an alkali with an acid as a way of producing fixed air (carbon dioxide). He also experimented with iron:

$$Fe + H_2SO_4 = FeSO4 + H^2$$

Iron plus sulphuric acid gives ferrous sulphate and pure hydrogen gas. Both of the above equations represent the reaction of an acid and a base (in the first case an alkali, in the second a metal) to produce a salt and a gas. It is the second reaction which led Cavendish to name "inflammable air", later christened hydrogen ("water maker") by Lavoisier. Others including Boyle and Priestley had noted this gas, but Cavendish was the first to describe its properties fully. Once again, he was confused by the famous phlogiston theory. He thought that his inflammable air WAS phlogiston as it was given off by the metal when attacked by the acid. He was wrong – the inflammable air came from the acid; but he had isolated hydrogen and could conduct further tests with this gas.

Cavendish followed up experiments which had been conducted by Priestley and his friend John Warltire, which led to the first understanding that water is a compound substance made from gases. The equipment consisted of sealed glass or copper vessels containing a mixture of hydrogen and air, fired by an electric spark. Everything was weighed before and after the experiments. All three scientists noted that after the explosion, the container walls immediately became dewy. Priestley thought that this was a result of condensation from the air, but Cavendish knew that it could not be so, because there was no change in the actual weight – in fact the weight of the water was equal to the combined weight of the hydrogen and oxygen used in the experiment. Meanwhile the weight of all the hydrogen and a fifth of the air was missing from the residual gases.

Cavendish further found that the number of measures of hydrogen was always twice the number of measures of oxygen used up, establishing that water contains these gases in the ratio 2:1. In these experiments he did not separate the oxygen from the air first, but he had already calculated that 20.8% of the common air consists of oxygen. Further experiments enabled him to establish that the rest of the air mainly consisted of nitrogen, which he called phlogisticated air. However he could never account for ALL the air in this way – he was left with about one part in 120 for which he could not account. His measures were indeed very accurate. Only 120 years afterwards was it found that, apart from trace gases, there is another constituent of the air, the inert gas argon, 0.93 or 1/107 by volume.

The most famous Cavendish experiment was also the most outlandish. In the outbuilding of his house on Clapham Common, he weighed the Earth. This experiment had in fact been devised by the Cambridge professor John Mitchell, who died before he could carry it out himself. It involved suspending two lead balls each two inches (5 cm) across from either end of a wooden beam, suspended by a wire from its mid-point. Two much larger lead balls weighing 350 pounds (155 kg) each were then suspended independently at either end of this contraption. The whole setup is called a torsion balance. Cavendish measured the amount of torsion or twisting in the beam caused by the gravitational attraction of the small balls to the larger, and multiplied this by the weight of the small balls. In this way he concluded that the density of the Earth is 5.48 times that of water. From this figure, knowing the dimensions of the Earth, he could calculated the weight of the world if he wanted to do so. The modern figure for the density of the Earth is 5.52 times the density of water, so Cavendish was very close. In more modern terms, what Cavendish was doing in this experiment was finding a value for G, the gravitational constant in Newton's formula, by measuring the faint pull of gravity of the large balls on the small ones.

As a footnote to this great man, the famous Cavendish Laboratory in Cambridge, subsequently midwife to so many Nobel prize laureates, was named after its benefactor, William Cavendish the seventh Duke of Devonshire, who came from a later generation of the same family, but who no doubt remembered that his remote uncle had been a gentleman scientist of no mean order.

Antoine Lavoisier (1743-1794) came from a wealthy middle-class family in France. Once described as "the spirit of accountancy raised to genius", he holds a unique place in the history of chemistry. He trained as a lawyer but became first a geologist, then a chemist. In 1768 he made what he thought was a good investment when he bought a share in a French tax farm. It turned out to be a very bad investment. However it gave him the privilege of collecting taxes, and if he collected more tax than the Government expected, he could keep the difference. The system of taxation in France at this time was notoriously unfair, as the nobility and higher clergy had obtained permanent exemptions, so the entire burden fell on the ordinary people, often in a most uneven manner. This did not go unnoticed, and when the French Revolution came in 1789, the reform of the tax system was a high priority, and the traditional "fermiers" – tax collectors – came into the firing line. However the job certainly provided Lavoisier with a good income, thought to be in the region of 150,000 livres a year (the equivalent of £13 million in today's money) at its peak (you begin to understand the problem).

Long before then, at the age of twenty-nine, Lavoisier unexpectedly married a girl of thirteen, Marie Paulze, daughter of one of his senior colleagues at the Ferme Generale; partly, it seems, to save her from the clutches of a lusting fifty-year old aristocrat. Though childless, the marriage was a happy one, and Marie became an accomplished chemist! There is a famous painting of the pair at their work by the most celebrated French painter of the day, Jacques-Louis David.

Lavoisier's most notable work is concerned with the process of combustion. He conducted his own experiments in which he weighed everything with the utmost care before and after combustion. He burnt metals and also let them rust. In both cases he found that weight had been added, not taken away, and the only place that the extra weight could have come from was the atmosphere; in fact the metals were adding oxygen. This effectively disproved the phlogiston model of combustion. So when Priestley burst in with Shelburne to announce his discovery of "dephlogisticated air", Lavoisier could only smile indulgently at the old-fashioned Englishman. He repeated Priestley's experiments as soon as the older man left to confirm his suspicion that this new air was nothing to do with phlogiston, but was a constituent part of the everyday atmosphere. He called it oxygen – "acid maker" – on the mistaken assumption that it is a constituent of all acids (hydrochloric acid, HCl, does not contain oxygen). Lavoisier's explanation was not accepted by Priestley, who stuck to phlogiston, recalling the later comment by the German physicist Max Planck: "A new scientific theory does not triumph by convincing its opponents and making them see the light, but rather because its opponents eventually die!" The naturalist Alexander von Humboldt put it another way – new ideas are at first denied, then considered unimportant, and finally are credited to the wrong person!

For his part, Priestley could be critical of Lavoisier and his carefully planned experiments. The trouble with that approach, he said, is that he never found anything unexpected – and it was Priestley who found all the new gases, not Lavoisier, who merely made sense of them.

To confirm the combustive power of oxygen, Lavoisier conducted an experiment so simple that one scratches ones' head to think why nobody had done this before. He placed a lighted candle on a float immersed in water, then placed an upturned jar over it, with its rim just under the surface of the water. The water level rose in the jar until it occupied one-fifth of its volume, and then the candle went out. It had simply used up the oxygen by the process of combustion. He also saw that the process of breathing in animals and people was another form of combustion, with oxygen being taken from the atmosphere and carbon dioxide put back into it. In short, Lavoisier not only solved the mystery of fire, but the whole process of combustion – combustion is oxidation.

Lavoisier set out his life's work in a book called the *Traité Élémentaire de Chimie* (*Elements of Chemistry*), first published in 1789 and quickly translated into other languages. This book and its predecessor, *Method of Chemical Nomenclature*, thoroughly modernized chemistry. Lavoisier replaced old terms, some of them dating back to antiquity, with a new nomenclature reflecting the actual chemicals involved. Oil of vitriol became sulphuric acid, and Epsom salts, magnesium sulphate. He introduced the idea of the balanced chemical equation, in the manner of a balanced account, where one side equals the other in terms of elements. This method had the advantage of predictive power. If, for example, hydrochloric acid is added to zinc, then zinc chloride forms, but the gas given off by this effervescent reaction must obviously be hydrogen.

In his book, Lavoisier also confirmed his idea of the conservation of matter. Chemicals react with one another to form new compounds, but nothing is actually destroyed by these reactions. Finally, the book included a list of all the known elements, thirty-three in all. Eight of these are now known to be compounds, and two were utterly wrong – light and heat ("caloric").

A busy man, Lavoisier took up other duties as a member of the Gunpowder Commission (to ensure the purity of gunpowder for the government) and as a deputy in the Third Estate, the new parliament meeting in 1789. However as a tax farmer, his card was marked, and he went to the guillotine in 1794, despite the frantic efforts of his wife Marie to save him. The judge was dismissive: "France has no need of chemists." He was fourth in the lists that day; his own father-in-law had been third. When a fellow member of the French Academy, the mathematician Legrange, hear the news, his comment was: "Only a minute to cut off that head, and a hundred years may not give us another like it."

## Chapter 6 – Later Georgian Science

It was during the eighteenth century that progress was made on that hitherto mystifying natural phenomenon, electricity, and this was due to the invention of devices which could actually provide a reliable supply of it for use in experiments. Static electricity of the type which can be generated by simply rubbing a balloon against a woolly jumper, and which produces alarming sparks, was known to the ancients. They realised that amber (for which the Greek word is *electron*) could be made to attract light objects when rubbed; but there is another and much more important form of it, electric current.

The first device capable of storing static electricity was the Leiden jar, invented by a Dutchman called Pieter van Musschenbroek (and independently by a German called Ewald von Kleist) in 1745. This is a glass jar, covered with insulating foil on both the outside and inside and provided with electrodes to gather and remove the electric charge. This device still needs a generator to make the electricity, generally provided by a friction wheel which could be wound around. The device proved surprisingly effective – indeed, Musschenbroek gave himself such a shock that he declared "that he would not take another for the kingdom of France"!

The theory of how the jar worked was provided by the American Benjamin Franklin (1706-1790), who recognised that the electricity has negative and positive charges, which he named, and flows in a current from one to the other. Franklin, one of the founding fathers of the United States, was a prolific polymath, inventor and all-round man of many parts. He served as the first ambassador of the United States in Paris in the years 1776-1785, that is, during the American War of Independence (1775-83) when France was a major ally of the new country.

Franklin hit on the idea that the lightning which flashes from the sky in big storms is in fact electricity – a giant spark from nature. To test this idea, he proposed flying a kite in a storm. His idea was taken up in France in 1752 where an iron mast 40 feet (12 m) high (rather than a kite, which would blow away) was set up, and proved Franklin right. Kite or mast, this was a dangerous activity, and a German professor called Georg Richmann was killed during one of his own experiments with a lightning rod the following year. Franklin, at one time President of the state of Pennsylvania, is remembered for his homely quotations, such as "An investment in knowledge pays the best interest!" Still feted in his native America, his is a familiar face from dollar banknotes, sporting a bald head and long hair; by his day, after a very long run, wigs were right out of fashion!

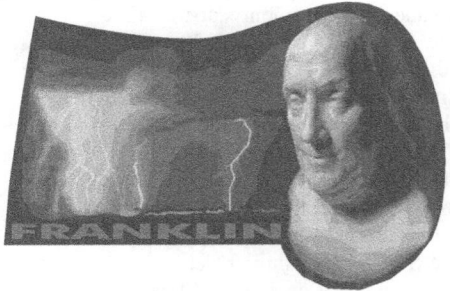

The real breakthrough in electricity came with the invention of the battery in Italy in the next generation. This arose out of a dispute between two Italian professors, Luigi Galvani and Alessandro Volta. (Galvani's name survives in the word galvanization, the process of providing a zinc coat to iron or steel vessels to prevent rusting.) Galvani had noted that the legs of dead frogs could be made to twitch, a phenomenon he associated with residual electricity in the body of the frog. Volta thought otherwise – he believed that the electric current was generated at the contact between the two metals, iron and brass, by which the frog's legs were suspended. Thus he set about experimenting with different sorts of metal plates, hoping to generate electricity.

Volta was already an electricity expert and he soon (1799) devised a brand new means of generating electricity, based on his hunch about the frog's legs. This consisted of a pile of silver and zinc discs, separated by cardboard disks soaked in brine. This produced a current of electricity when the top and bottom were connected by a wire. This was a dramatic advance. The Leiden jar released its whole charge in one go, but the voltaic pile, as it was called, produced a steady flow of electricity which could be – and very quickly was – switched on and off in experiments.

The first tentative steps towards an understanding of the process of photosynthesis were also made around this time. This was the work of the Dutchman Jan Ingenhousz (1730-1799). Pushing all kinds of green leaves into glass jars, immersed in water, he noted that when left in the shade, nothing happened. When placed in the sun, however, every different kind of leaf gave off bubbles of a gas, which he identified as oxygen. He also placed his jars by the fire, when nothing happened, so demonstrating that it was light from the sun, and not heat, which made the process work.

There were also futher advances in astronomy at this time. The planet Uranus was discovered by William Herschel and his sister Caroline in 1781. Naturally this caused a sensation as it was the first planet which had not been known to the ancients. Also around this same time, the West Yorkshireman and Cambridge professor John Mitchell (1724-93), author of the Cavendish experiment with the lead balls, established the concept of black holes. It was known from the work of Romer in the previous century that light has a finite speed, and Mitchell reasoned that no light could escape from any body of 500 or more times the diameter of the Sun, and of similar density. This is a remarkably modern concept, implying as it does that Mitchell knew (as Einstein established in 1915) that light, despite having no apparent mass (part of the Newtonian equation) is influenced by gravity. Mitchell also reasoned that we would only be able to infer the presence of black holes from the motions of bright bodies around them – just, in fact, as happens today.

At the end of the eighteenth century there was increasing interest in the subject of heat, and the nagging feeling that the current theory, based on the principle of caloric, might be wrong. Caloric, as explained by Lavoisier and his predecessors, was a weightless substance – like light – which flowed from any body when it was warmer than its own environment, from warm to cool. This meant that any body could only contain a certain amount of caloric, which would eventually run out. The man who disproved this idea was not a scientist at all, but an American adventurer, Benjamin Thompson (1753-1814), who came from Massachusetts. Having taken the losing side – Loyalist, or British – in the American Civil War, Thompson sought work in Europe, and he found it in the most surprising place – Bavaria. Working with the military in the government service, he was such a success here that he was created Count Rumford in 1791. One of his achievements was the design and construction of the famous English Garden in Munich.

It was the process of boring out the barrels of cannon from solid metal which convinced Thompson that heat was something to do with motion, and nothing to do with caloric, because the supply of heat given out by the friction of the drill bit against any lump of metal proved to be inexhaustible. The drill bit was turned by horses, and in an experiment written up in 1798, Thompson proved it was quite possible to boil water using the heat generated from this process. If the drill bit was blunt enough, this process could continue indefinitely – more and more water could be boiled. It was literally horse-power, the pulling power of moving horses, which created the heat. This idea, slow to catch on at first, underlay the subsequent theory of the conservation of energy. The hay put into the horses (chemical energy) became kinetic (moving) energy in the drill, to be dissipated as heat from the cannon and the bit.

In 1798 Thompson (now Rumford) returned to England, where he did something rather unexpected (given his rather unsavory reputation for skullduggery in the American War). After discussions with Henry Cavendish and Joseph Banks, he founded a brand new institute, the Royal Institution, meant to be a museum, research and educational establishment. He raised the necessary funds through public subscription, obtained premises and appointed a lecturer to give public demonstrations and lectures on the new sciences. The RI opened its doors in 1800, and it was to prove a great success, particularly under its second lecturer, Humphry Davy, and his later assistant, Michael Faraday. Rumford himself soon disappeared into Europe, where he was briefly married to Lavoisier's widow. The RI, still going strong today, played an important role in disseminating new knowledge, particularly of chemistry, often seen as the province of industrialists – coal and potash miners and dyers – rather than of the gentlemen, who preferred to dabble in geology and natural history (as did, for example, Charles Darwin).

Humphry Davy (1778-1829), born to a humble family in Cornwall, was the first scientist to take advantage of the new opportunities offered by Volta's battery. Davy received no more formal education than was available at Truro grammar school. The French Revolution, which began in 1789, had turned into the Reign of Terror by 1793-4 (and also war with England by 1794), sending a flood of refugees to all corners. Davy befriended one of these, a French priest, who taught him French, and so it was that Davy was able to read Lavoisier's *Traite Elementaire* in the original French at the end of 1797. He also befriended Gregory Watt, the son of James Watt, who had been sent to Cornwall to try to recuperate from tuberculosis in the relatively mild climate. He in turn had studied chemistry at Glasgow University. So it was that Davy picked up enough chemistry to obtain a job as an assistant to Thomas Beddoes of Bristol. This man was experimenting with gases to try to find a cure for tuberculosis, and had set up a clinic in Bristol.

Davy experimented with various gases in his new job, and nearly killed himself with carbon monoxide poisoning, but the gas which made his name was laughing gas – nitrous oxide – which made him happy! He published a book on this gas in 1800. He then became interested in the possibilities of the new voltaic pile, having convinced himself that there was a significant relationship between electricity and chemistry. He was right. Meanwhile the first lecturer at the newly formed Royal Institution was not proving a success in his second year, so Rumford looked for a replacement. He found Davy, by now the brightest young star in British chemistry. He took up his appointment in 1801 with a salary of 100 guineas a year, plus accommodation at the RI. By 1802 and still only 23 years old, Davy was the professor at the RI. Here he soon established a fine reputation, and became a highly paid guest lecturer elsewhere, especially on important subjects such as the role of chemistry in agriculture.

The new batteries enabled the technique of electrolysis for the first time. Two electrodes would be placed in a bath of chemicals, and a current allowed to flow through them. Davy's most successful results were with the molten salts of alkali metals, which separated into their different fractions, the highly reactive metal accumulating at the negative electrode (cathode). Passing current through potash and then soda, Davy discovered two previously unknown metals, which he called potassium and sodium. These findings caused a sensation at the time, similar to the discovery of phosphorus half a century earlier, given the spectacular behaviour of the newly discovered elements. A small quantity of potassium dropped into water immediately burst into flame and whizzed about on the surface of the water issuing a fierce hiss. It was in fact sucking oxygen from the water, liberating hydrogen which ignited because of the heat generated by the reaction. Davy is said to have danced around his laboratory in glee when he first tried this – he knew he had found a spectacular new element.

He also isolated chlorine from a solution of brine in this way:

$$2NaCl + 2H_2O = 2NaOH + H_2 + Cl_2$$

In addition, he established that hydrogen, rather than oxygen, is the essential element contained in all acids. He did this by combining hydrochloric acid with potassium and observing that this produced potassium chloride and hydrogen gas, but no oxygen. Then using relatively common, known minerals which were already suspected to contain unknown elements – lime, magnesia, strontia and baryta – he duly isolated calcium, magnesium, strontium and barium.

It is now known that the crust of the earth is composed overwhelmingly (98.2%) of just eight elements: in order, oxygen, silicon, aluminium, iron, calcium, sodium, potassium, and magnesium. So in his short career, Davy had discovered four of these, and another, oxygen, had only been isolated in the previous generation.

In 1812 Davy was knighted, and shortly afterwards married a wealthy widow. An extensive tour of Europe soon followed; Davy took his assistant Michael Faraday with him. On his return to England, Davy began his research to solve a long-outstanding industrial problem, that of gas in mines, which had killed countless miners over the generations, either by suffocation or explosion. This produced the first successful miner's safety lamp, which still bears Davy's name. However, the research which led to the design had clearly been so meticulous – for the lamp is not a simple thing, and the problem had foiled every previous attempt – that is has long been suspected that Faraday must have had a great deal of input to the process. This may be unfair, but Davy had the reputation as a somewhat slapdash scientist. After this time Davy achieved little of scientific note, but became such a snob that he was the only fellow of the RI to oppose the election of Faraday to his own fellowship. He died in 1829 aged only fifty, thought finally to have overdone things with the laughing gas which he himself had popularized, and to which he had become addicted.

\*\*\*\*\*\*\*\*\*\*\*\*\*\*\*\*\*\*\*\*\*\*\*\*\*\*\*\*\*\*\*\*\*\*\*\*\*\*\*\*\*\*

James Watt's steam engines soon began to find new applications. The first man to produce a steam locomotive was the Cornishman Richard Trevithick (1772-1833), whose invention dates from 1804 when he installed a locomotive at an ironworks in Wales. This ran for only three trips before being abandoned, but Trevithick made improvements. A more successful model was produced for a colliery on Tyneside in the north-east of England, where it was noticed by a young George Stephenson (1781-1848). This self-taught Geordie engineer went on to produce the two most famous early locomotives. These were the *Locomotion*, which ran on the first railway, the Stockton to Darlington, from 1825, and the *Rocket*, which ran on the Liverpool to Manchester railway from 1829. The *Rocket* was far in advance of any previous locomotive and can be counted as the first truly successful railway locomotive.

\*\*\*\*\*\*\*\*\*\*\*\*\*\*\*\*\*\*\*\*\*\*\*\*\*\*\*\*\*\*\*\*\*\*\*\*\*\*\*\*\*\*\*\*\*\*\*\*\*

Another one-time lecturer at the Royal Institution was Thomas Young (1773-1829), known for his sheer brilliance whilst at Cambridge as "Phenomenon" Young. He took a turn at the RI in the years 1801-3, so overlapping with the early Davy, but his lectures were not a success – they flew too high over the heads of the audience.

Young was born into money – the son of a banker – at Milverton in Somerset. He was a child prodigy, for example learning many languages before he was ten years old. At the very early age of 21 he was elected a fellow of the Royal Society, this for a paper explaining the focusing mechanism of the eye – the way the muscles change the shape of the lens. Under the influence of his great uncle, one Richard Brocklesbury, he trained for a medical career, and when Brocklesbury died, Young, by now aged 27, went to London to take over his house and medical practice in 1800. When there he became one of the few people to make progress in the deciphering of the Rosetta Stone. He was also the first person to appreciate that colour vision is produced by combinations of three primary colours, red, blue and green, which affect different receptors within the eye. However he is best remembered for his experiments on light, which showed that it travels as a wave, rather than, as Newton thought, a stream of weightless particles. His idea, following on from Christiaan Huygens, was that different colours of light correspond to different wavelengths. Using Newton's own experimental data, he actually calculated the wavelength of red light as $6.5 \times 10^{-7}$ metres, which agrees well with modern estimates.

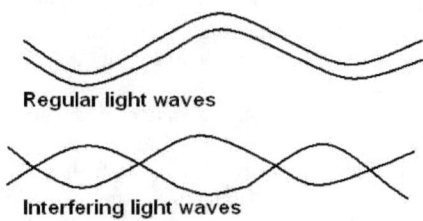

**Regular light waves**

**Interfering light waves**

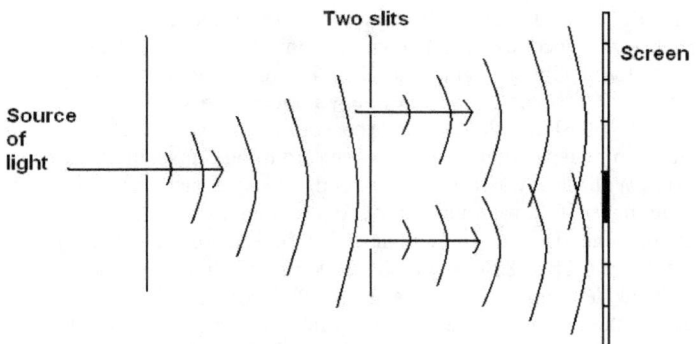

Young's double-slit light experiment

In 1801 he announced a key contribution to what was to be a long-lasting debate: the idea of interference in light waves. When a light is shone on a screen, then a single colour may be observed, but if that light source is split into two, the waves may interfere with one another (see diagram). If the "ups" and the "downs" overlap – that is, the waves are out of phase – the result can be a dark strip, that is, shining a light can produce darkness! This was demonstrated in Young's famous double-slit experiment. A light is shone through a slit in a card, then passing to another card with two slits in it. This second card divides the light, which is then shone on a screen. The resulting pattern on the screen is made up of distinct bands of variable colour, some the original colour, but others, where the two beams intersect, of darker or even completely black shades. This process – the changes in intensity of light or other waves after passing through a narrow aperture – is known as diffraction. If light were just a stream of particles, this would not happen – it would pass through the two slits and reach the screen in two widely separated lines, with nothing in between. So Young had designed a simple experiment, which as he said, anyone could carry out without special equipment, which proved beyond any reasonable doubt that light travels as a wave.

Newton's influence was still so strong that despite the evidence, Young was not believed – how could adding two beams of light together produce darkness? However, in 1817, his work was corroborated by the French engineer Augustin Fresnel, who designed a similar experiment. Fresnel is probably better-known for designing a special lens for use in lighthouses.

Meanwhile a new instrument became available for those wishing to study light – the spectroscope, a combination of a prism (or other mechanism for splitting light into its different colours) and a microscope. The first person to achieve any noticeable results with this was William Wollaston (1766-1828), who noticed that there are many distinct lines in the spectrum of light coming from the Sun, some light, some dark. Wollaston was a prolific scientist and we shall meet him again. He never followed up his discovery of the spectral lines. However someone else did – a German industrial scientist called Josef von Frauenhofer (1787-1826), who did not do this for fun – he worked in the optical laboratory of the Munich Philosophical Instrument Company, a manufacturer of lenses, for which Germany was soon to establish a worldwide renown. In 1814 Frauenhofer saw lines in the spectrum of light from the Sun – many, many of them, in fact 576 in the wavelengths of visible light. These lines were also visible in the light coming from the planets and the stars. Frauenhofer did not know what caused these lines to be there, but it was he who set the hounds on the trail, and for that reason the lines are called Frauenhofer lines, and not Wollaston lines.

*******************************************

Another scientist of this era originally from the English backwoods is John Dalton (1766-1844), a man famous partly for being colour blind, a condition which is called Daltonism after him. He was born into a Quaker household at Eaglesfield on the edge of the English Lake District, but spent most of his working life in the rapidly expanding industrial city of Manchester, where he worked as a teacher and private tutor.

In 1788 a French chemist called Jean-Louis Proust had discovered an important property of gases and other compounds – they always combine in definite proportions by weight. This might be 2:1 or 3:1, but it was never, for example, 3.5:1. Dalton came to the conclusion that the reason for this must be that each element is in fact made up of a vast number of tiny particles, which he called atoms. If the particles of one element weigh three times as much as the other then the ratio of the weights is bound to be exactly 3:1 as Proust had observed. Dalton soon concluded that hydrogen is the lightest element, and he gave this an atomic weight of 1. He knew that water is a combination of hydrogen and oxygen, in proportion by weight of 1:8. Therefore he gave oxygen an atomic weight of 8. In fact he was mistaken at this point, because water is a combination of two hydrogen atoms for every one of oxygen, and the real atomic weight of oxygen is therefore 16.

The subsequent discovery that a single element can have different isotopes (of slightly different weights) means that in fact compounds are not always found in perfect proportions of elements by weight. Nevertheless Dalton's theory of atoms went beyond the mere speculations of Democritus and Gassendi and in fact was to prove the basis of future progress in both chemistry and physics. It was published in his seminal work, *A New System of Chemical Philosophy* (1808), which found a wide audience in the civilized world. Dalton also established that his atoms are practically indestructible, and that it is to all intents and purposes impossible either to create or destroy say an atom of hydrogen.

Dalton himself shunned honours and preferred to live out his simple Quaker life in Manchester, but he was loaded with honours in any case. He died at the age of 77. Although he expressed a wish for a simple Quaker funeral, the event drew forty thousand people and a hundred carriages – Dalton was the original local hero.

\*\*\*\*\*\*\*\*\*\*\*\*\*\*\*\*\*\*\*\*\*\*\*\*\*\*\*\*\*\*\*\*\*\*\*

Further investigations on the nature of gases threw more light on the idea of molecules, whereby atoms of the same element combine to form the actual gas which occurs in nature. As a gas, hydrogen is $H_2$, oxygen $O_2$ and nitrogen $N_2$. In 1808 the French chemist Gay-Lussac (1778-1850) found that when gases react to form a new gas, the resulting volume can be expressed in simple numbers. For example, when two volumes of hydrogen combine with one volume of oxygen, the result is two volumes of water vapour. In practice this means that:

2 molecules of hydrogen + 1 molecule of oxygen = 1 molecule of water vapor.

In 1811, the Italian Amadeo Avagadro concluded from this that equal volumes of gas contain equal numbers of molecules (at constant temperature and pressure), whatever the gas is – in other words, no matter how big the individual gas molecules are.

As an example, three cubic metres of hydrogen ($H_2$) combine with one cubic metre of nitrogen ($N_2$) to give two cubic metres of ammonia ($NH_3$). Let us assume that one cubic metre of hydrogen contains a thousand hydrogen molecules. Therefore 3 cubic metres contains 3000 molecules, that is, 6000 hydrogen atoms altogether. One cubic metre of nitrogen also contains 1000 nitrogen molecules, that is, 2000 nitrogen atoms. Put these two together and we obtain 2000 molecules of ammonia – one for each nitrogen atom, and one for every three hydrogen atoms. In every case there are one thousand molecules per cubic metre.

With the water vapour example, 2000 molecules of hydrogen give 4000 atoms. These combine with 1000 molecules of oxygen, or 2000 oxygen atoms. The result is 2000 molecules of water vapour $H_2O$. Once again this remarkable result is obtained – the same volume of gas always contains the same number of molecules. It was this discovery which led to the realization that gases exist in combinations of their own atoms – that hydrogen gas is $H_2$, and not just H.

It was to be many years before Avagadro's law was accepted. However, he is also famous for his number, which is the number of molecules in a cubic centimetre (technically in a "mole") of gas. This is rather confusing, because Avagadro did not establish this number. Rather, it was named after him when it was eventually discovered! In other words, Avagadro's law states that there is always the same number of gas molecules in the same volume, no matter what the gas – but what is the number? This value, not established until the twentieth century, is very large indeed:

$6.022 \times 10^{23}$ per mole

Gay-Lussac (in 1802) and Dalton (in 1801) also independently enunciated another law of gases, the law of volumes, or Charles' Law (named after the Frenchman Jacques Charles, thought to have discovered it in 1780). This states that at constant pressure, the volume of a gas increases or decreases by the same factor as temperature (on the absolute scale), so if the temperature rises by say 10%, so does the volume. In fact, in 1802 nobody knew what absolute zero was (this was later established by Lord Kelvin as -273 degrees Celsius), but Gay-Lussac's estimates were close to that figure at -266.6 degrees. (At such low temperatures, gases have liquefied or frozen.)

Incidentally, Gay-Lussac was apparently a man of strong mettle. In 1804 he decided to try out some experiments in a hot air balloon with a colleague called Biot. Apparently the ascent up to 4,000 metres/13,000 feet went reasonably well but the descent was another matter. Towards the end of it Biot lost all possession of himself, and it terminated in an inglorious though not fatal crash landing.

\*\*\*\*\*\*\*\*\*\*\*\*\*\*\*\*\*\*\*\*\*\*\*\*\*\*\*\*\*\*\*\*\*\*\*\*\*\*

Dalton made an attempt at establishing a table of elements, but he made limited progress, hampered as he was by insisting on the use of clumsy, home-made equipment, and also by his colour blindness (he could not distinguish pink from blue). The next famous chemist, the Swede Jons Berzelius (1779-1848) made much more progress. He was another student who turned aside from medicine, at which he had performed poorly, for a much more fruitful career in the new science. Like Davy he made much use of voltaic piles to study the elements. For example, with copper sulphate, he found that copper would accumulate at the (negative) cathode. He realised that this must mean that the copper part of the copper sulphate carries a positive electric charge, as it is attracted to the negative electrode. This led him to propose a new and far-reaching theory, that all compounds consist of positively and negatively charged parts.

He then embarked upon an exhaustive study of the elements and compounds known at the time. Following on from Dalton, by 1818 he had established the atomic weights of 45 of the 49 accepted elements, and he had analysed over 2000 compounds. He found, however, that his new theory did not work in every case – in particular, it did not work at all with the large range of organic compounds. As early as 1807, Berzelius formalised the long-recognised distinction between what he now called inorganic and organic chemicals. Inorganic chemicals including water, iron and even salt can be heated and change their condition, evaporating, glowing red hot or melting for example, but then revert to their original condition when cooled. However a lump of wood, once burned, can never revert to being a lump of wood.

It soon came to be acknowledged that organic compounds are much more complex than inorganic ones, and all of them are based around one element, carbon. Berzelius came to the conclusion that they are held together by a non-electrical mechanism which he called "life force". This new idea, known as vitalism, was to have a long run in front of it, in the manner of phlogiston, despite the fact that, like phlogiston, the life force never showed up in any experiment. It was another blind alley. What Berzelius had in fact discovered was only one form of chemical bond, the ionic bond, by which for example a positively-charged sodium cation combines with a negatively-charged chlorine anion to make salt. Organic compounds use other forms of bond to stick together, notably covalent bonds, as discussed in chapter 12.

However Berzelius was far from finished. Whilst Lavoisier had established a framework for chemistry, it was the Swede who provided the bricks. For example the chemical elements were known by local names in every country – in German, hydrogen had been christened "Wasserstoff" – water stuff. Berzelius proposed that a Latinized name be used in scientific papers, and then came up with a scheme for a symbolic chemical alphabet. Every element would from now on be known by its initial, and then another letter if required. Hydrogen became H, oxygen O, gold Au (Latin aurum), silver Ag (Latin argentum) and iron Fe (Latin ferrum). Compounds would be written in this new script, so that for example carbon monoxide became CO. If more than one atom was involved in a compound, this would be indicated by a subscript, so that carbon dioxide became $CO_2$. Ammonia, which contains one nitrogen atom and three of hydrogen, became $NH_3$. The new mathematically-based formulae could indicate not only what chemicals were involved in a reaction, but their relative quantities. Lavoisier's nomenclature was this:

Zinc + hydrochloric acid = zinc chloride + hydrogen

This became:

$Zn + 2HCl = ZnCl_2 + H_2$

Berzelius is also credited with the discovery of four new elements – silicon, thorium, selenium and cerium. The last of these is the commonest of the so-called rare earths, famous for being neither rare (cerium is as common as copper) or earths (they are metals). The rare earths fill a full seventeen places in the periodic table; most of their names are completely unfamiliar, but many have found modern industrial applications. Europium, for example, is a constituent of the inks which are used to print Euro bank notes. The series comprises the group of fifteen metals known as the lanthanides plus scandium and yttrium. Many were first discovered in mines in Sweden, notably the one at Ytterby. They were discovered over a long period – a hundred or more years, starting in the 1790s – by painfully slow experimentation. An ore would be dissolved in acids, forming a solution of different salts. These would gradually crystallize out in turn. This process would be repeated thousands of times if necessary until the new element revealed itself. It was a dull grind, undertaken by a whole series of conscientious Swedish chemists, including Berzelius, but it worked.

During the course of the nineteenth century, new chemical elements were discovered at frequent intervals. The heavy metal platinum was first properly investigated in the middle of the eighteenth century, when samples of its ore, a substance called platina, were dredged from South American rivers and sent to Spain. It proved extraordinarily difficult to refine the platinum in a malleable form, which could be used to make artifacts, though this feat was eventually achieved by a Frenchman called Chabaneau, working for the King of Spain. Hence platinum is completely different to gold, having a very short history as a valuable metal. Chabaneau's difficulties were due to presence of other, at that time unknown metals in the platina. These were refined out in the early nineteenth century by two Englishmen. Of these, the Yorkshireman Smithson Tennant (1761-1815) isolated iridium and osmium. The other man, William Wollaston, was the son of a Norfolk rector who, as noted above, was the first person to observe what are now called Frauenhofer lines. His researches into the platina ore led to the discovery of two more new elements, rhodium and palladium. However it was the process which he invented to refine platina into malleable platinum which made his fortune, because a strong demand for the metal arose from the gunsmith trade. It was used to make the contact points in flintlock pistols. The fashion to use it as a jewelry metal is a later phenomenon. The five metals – platinum, iridium, osmium, rhodium and palladium, plus a sixth, ruthenium, are known as the platinum group. They form a distinct cluster in the periodic table, and they have similar physical and chemical properties. Ruthenium is much the rarest of these metals and it was not discovered until 1844 by the German, Karl Claus.

In 1817 the German Friedrich Stromeyer isolated cadmium from the zinc oxides with which it is normally associated. The sulphides of cadmium were soon found to yield a brilliant range of pigments, in every colour except blue, varying according to impurities. These then became available to the painters of the nineteenth century, giving groups such as the Pre-Raphaelites and the Impressions a range of brilliant colours simply unavailable to earlier generations of painters.

In 1827, a German called Friedrich Woehler succeeded in isolating metallic aluminium (in fact the commonest of all metals in the crust of the Earth). This process proved so difficult that for a few decades, aluminium became more valuable than gold. The French Emperor Napoleon III ordered a set of cutlery to be made from it, no doubt to entertain the crowned heads of Europe, though today it would be scarcely thought fit for a school canteen. Woehler also succeeded in synthesizing urea from non-living materials. As this is one of the constituents of urine, it is a product of living matter. The creation of organic matter from inorganic should have discredited the life-force theory, but it failed to do so.

About the same time (1829), another German chemist, Johan Döbereiner, noticed that the newly-discovered element bromine had properties which seemed to place it midway between chlorine and iodine. (These elements are now called the halogens or salt-makers.) Also its atomic weight lay exactly half-way between the two. He then found that two other groups of elements followed the same pattern – strontium, calcium and barium; and selenium, sulfur and tellurium. Döbereiner called these groups triads, but could only link nine of the 54 known elements in this way. His contemporaries dismissed the idea as mere coincidence, but it was no coincidence. It was the first glimpse of the periodic table of chemical elements.

*******************************************

One notable astronomical discovery dates from the middle of the nineteenth century, in fact from 1846 – a brand new planet, Neptune, the fourth-largest body in the solar system. Several astronomers and mathematicians had noted discrepancies in the orbit of the next planet, Uranus, for which the best explanation was the gravitational attraction of a planet orbiting outside of it. Credit for the discovery is usually accorded to the Frenchman Urbain Le Verrier, who not only predicted the existence of the planet, but told the astronomers at the Berlin observatory more or less exactly where to look for it.

## Chapter 7 – The First Geologists

The establishment of geology as a new science really begins with the publication of *The Theory of the Earth* by James Hutton (1726-1797), which appeared in 1788 and in its full version in 1794. This man was based in Berwickshire, in the south-eastern corner of Scotland, where he inherited a set of stony fields scarcely fit to be called a farm. Until the time of Hutton, as noted, it was widely accepted that the Earth was created in 4004 BC. Hutton realised that this could not possibly be so. One illustration he used was the network of Roman roads. In England at least, the Romans had gone by 410 AD, but where they had not been buried, large sections of the roads which they had built of stone flags and setts remained visible on the ground. If erosion had so little impact in the 1400 years or so since these roads were built, how could the whole word have been created a mere four and a half thousand years before that? Erosion would not wear away a single stone of the roadway in that time.

However the structure which really convinced Hutton was what is now called an unconformity, where newer rocks lie on top of older rocks, often dipping into the ground at completely different angles, with a clear and distinct break between them. He found such a site a few miles west of St Abb's Head on the coast near Berwick called Siccar Point, which became one of the most famous sites in geology. The unconformity lay between vertical beds of Silurian slates and grits, and horizontal beds of Devonian sandstones which had been deposited on top of them millions of years afterwards (in fact both Silurian and Devonian rocks are over 400 million years old). Ever since then the heavy boots of structural geologists, geochronologists, geochemists and paleomagneticists have beaten a trail to the famous Siccar Point. Remote places like this may as well not exist for the general public, but for geologists they are the centre of the universe.

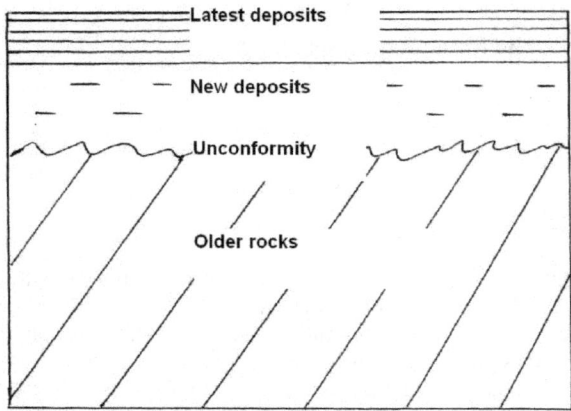

An unconformity, as is found at Siccar Point in Scotland, where vertical beds of Silurian slates and grits are overlain by horizontal beds of the newer Devonian sandstones

Hutton saw that the earth's crust was recycling itself, and that the constant process of erosion of the land into the sea meant that land must also rise by uplift, that indeed the sea floor could form mountains in the next cycle. In his famous words, he saw "no vestige of a beginning, no prospect of an end." He was the first pioneer of the concept of deep – geological – time; but he had no idea how deep. He also saw no need to invoke catastrophic processes in the creation of the landscape as he saw it. If enough time were allowed, processes which could be seen operating today, including normal erosion and vulcanism, could explain everything: *the present is the key to the past.* However, Hutton's books proved so impenetrable that they had relatively little immediate impact, but a more accessible version was published within a few years by one John Playfair, entitled *Illustrations of the Huttonian Theory of the Earth* (1802). This book did reach a wider audience.

In the next generation after Hutton great strides were made in the establishment of the sequences in the English stratigraphical column and the fossils it contains. The most important name from this era is William Smith (1769-1839) (born in the same year as Georges Cuvier, Napoleon and Wellington). A man of humble origins – his father had been a blacksmith – Smith worked as a surveyor of canals and coal mines in the Somerset area. In 1799 he published the world's first real geological map, of the area around the city of Bath. By dint of much hard labour and often with little money, he published the first geological map of England, Wales and the southern part of Scotland in 1815. In some ways, however, it was an unfortunate profession, as his wife was mentally unstable – described in fact as mad and bad – exhibiting nymphomaniac behaviour. Smith eventually overstretched himself financially and was obliged to spend six weeks in a debtors prison.

Smith was the first person truly to appreciate the significance of fossils. Up until his time, they had been collected – indeed, avidly collected – as interesting curiosities. Smith realised they were much more than this. They act as clocks embedded in the rocks, and can be used to identify sedimentary rocks definitively. Especially in the Jurassic sequences of southern England, there are repeated formations of sandstones, marls, clays and so on – often only a few feet thick – which look very similar. One sandstone could look just the same as another 150 metres (500 feet) higher in the sequence – but they would contain different fossils, and so could not be the same rock. Conversely, if a fossil assemblage found in Dorset matched another found in Yorkshire, it must be the same rock. Smith saw that the same succession of fossil groups from older to younger rocks could be found all over southern and eastern England.

The 1815 map used symbols to mark canals, tunnels, tramways and roads, collieries, lead, copper and tin mines, together with salt and alum works. (Alum is a term for aluminium salts, used for purifying water and other cleansing.) The various geological formations were indicated by different colours; the maps were hand coloured. The map is remarkably similar to modern geological maps of England. A copy of it still hangs at the Geological Society in London.

The Jurassic (250-200 million years ago) was the first great period for the fossil collectors. In 1825 a country doctor called Gideon Mantell published details of a Jurassic *Iguanodon* found by his wife poking out of the Sussex Weald. The year before, William Buckland had named a another Jurassic monster, *Megalosaurus*. Buckland (1784-1856) was one of the leading geologists of his day, and is also credited with the discovery of the Red Lady of Paviland (actually a man) in a cave in Wales, the oldest modern human bones found in Britain, dated to 33,000 years ago.

The best-known of the Jurassic fossil hunters was actually a woman, Mary Anning (1799–1847), who became famous for a number of important finds she made in the Jurassic marine fossil beds at Lyme Regis on the coast of Dorset where she lived. The world watched in amazement as she dug out previously undreamed-of, long extinct creatures, whose specimens fetched fancy princes from aristocratic collectors.

Mary worked on the Blue Lias cliffs made up of thinly bedded shales. Her discoveries included fantastic specimens of ichthyosaurs, plesiosaurs (large marine reptiles with four flippers and a long neck), pterodactyls and some important fish fossils. Mary herself was an uneducated seaside woman, and was never going to be able fully to participate in the scientific community of nineteenth-century England – dominated as it was by wealthy Anglican gentlemen. She struggled financially for much of her life. Her family was poor; her father, a cabinetmaker, died when she was only eleven. Nevertheless she remains well-known to this day.

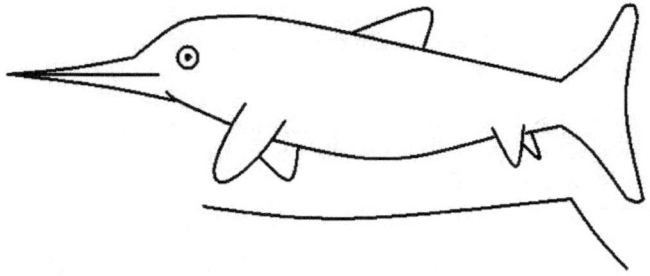

Ichy the ichthyosaur and below, the line of its spine

The real founding publication of geology came to be seen as the work of Charles Lyell (1797-1875), thirty years later. Brought up in the New Forest not far from Southampton, Lyell came from a wealthy family and had been expected to pursue a career in the law – always it seems that men destined for law, medicine or the church by their families were turning aside into the evolving sciences! He became greatly interested in geology, and neglected the law, at first to the chagrin of his father; but his father did eventually come on board. Lyell made a European tour in 1828, partly in the company of another famous geologist, Sir Roderick Murchison (1792-1881). It was what Lyell saw in Sicily, in the area of the volcanic Mount Etna, which convinced him that the Earth had been formed by the same processes which operate today, but operating over immense ages of time. He was especially impressed with the sheer size of the volcano, 90 miles (150 km) across at its base – how many lava flows must it have taken, and over how long, to build a structure of that size?

He codified much of modern geology as it is understood today when his *Principles of Geology* appeared in three volumes in the years 1830-3. This was one of the most influential books taken by Charles Darwin on his voyage on the Beagle, which began two years later. Lyell developed Hutton's idea that "the present is the key to the past", the idea being that geological remains from the distant past can, and should, be explained by reference to geological processes now in operation and thus directly observable. Lyell interpreted geological features as the result of the steady accumulation of minute changes over enormously long spans of time. This is called Uniformitarianism, the assumption that the same natural laws and processes that operate now have always operated in the past, and function at similar rates. The term *uniformitarianism* itself was coined by William Whewell, who also coined the term *catastrophism* for the idea that the Earth was shaped by a series of sudden, short-lived, violent events (or Neptunism, in which most of geology was put down to Biblical-style floods).

A chair of geology was established at King's College, London in 1831, and Lyell was appointed professor. However, he gave up the job only two years later, finding its duties irksome. By now he had an income from his books, and he produced frequent new editions to keep pace with this rapidly evolving subject. His idea of uniformitarianism has remained a key principle of geology, but modern geologists, while accepting that geological processes have operated across deep time, no longer hold to a strict gradualism. There can be catastrophic events as well – most famously, large meteors crashing into the earth. Charles Darwin, who was a geologist before he was a biologist, was to apply a similarly uniformitarian view to evolution.

Geologists today apply the principle of uniformitarianism rather as other scientists apply the principle of parsimony, also mysteriously known as Occam's Razor. This states that a simple explanation is always to be preferred to a complicated one – it is the most "parsimonious". When a new theory comes along in geology, the more it invokes processes which cannot be seen operating on the earth today, the more scepticism it provokes. An example is the modern idea of the Snowball Earth, which proposes sea ice at the equator.

*Principles of Geology* was first published shortly before another major breakthrough in the field. In 1837 a Swiss academic called Louis Agassiz (1807-1873) became the first scientist to propose that the Earth had been subject to a past ice age. His attention had been drawn to the presence of erratic boulders, rocks far away from their original outcrop, by one Jean de Charpentier. Several walkers had by then arrived at the conclusion that the erratic blocks of alpine rocks scattered over the slopes and summits of the Jura Mountains – which contain no glaciers – must have been moved there by ice. In 1840 Agassiz published a work in two volumes entitled *Études sur les glaciers* ("Study on Glaciers"). In it he discussed the movements of the glaciers, their deposits (moraines and erratics), their influence in grooving and rounding the rocks over which they travelled, and in producing the striations and *roches moutonnées* (boulders which are smooth on the upstream side, but plucked out on the lee side) seen in Alpine-style landscapes. He not only accepted Charpentier's idea that some of the alpine glaciers had extended across the wide plains and valleys drained by the Aar and the Rhône, but he went still farther. He concluded that, in the relatively recent past, Switzerland had been another Greenland. Instead of a few glaciers, one vast sheet of ice, originating in the higher Alps, had extended over the entire valley of northwestern Switzerland until it reached the southern slopes of the Jura, which, though they checked and deflected its further extension, did not prevent the ice from reaching the summits of the range.

Mont Blanc in the heart of Agassiz territory

Agassiz travelled to Scotland with the English geologist William Buckland, in 1840. The two found in different locations clear evidence of ancient glacial action. The mountainous districts of England, Wales and Ireland were also considered to constitute centres for the dispersion of glacial debris; and Agassiz remarked "that great sheets of ice, resembling those now existing in Greenland, once covered all the countries in which unstratified gravel (boulder clay shunted by ice) is found." Originally an old-fashioned Noachian Flood proponent, Buckland became a firm supporter of Louis Agassiz and his theory of glaciation.

\*\*\*\*\*\*\*\*\*\*\*\*\*\*\*\*\*\*\*\*\*\*\*\*\*\*\*\*\*\*\*\*\*\*\*\*\*\*

Nineteenth-century geology was littered with bone-shaking arguments amongst its early proponents. One of these was a notable dispute between the aristocratic Sir Roderick Murchison, and his former friend, Cambridge professor Adam Sedgwick (1785-1873), who grew up on a humble farm in Yorkshire.

In 1835 Murchison and Sedgwick presented a joint paper, under the title *On the Silurian and Cambrian Systems, Exhibiting the Order in which the Older Sedimentary Strata Succeed each other in England and Wales*. This was the foundation of the modern Palaeozoic timescale (for rocks between 542 and 250 million years old). When traced away from its source area in south Wales, Murchison's Silurian series came to overlap Sedgwick's Cambrian sequence, provoking furious disagreements. Sedgwick watched appalled as Murchison's Silurian colours crept up the map of Wales, reducing his Cambrian sequence to a rump. The dispute also spread to English formations. Sedgwick later wrote to Murchison:

"Your nomenclature of the older English rocks is false for the simple reason that below your true Silurian groups (beginning with the Wenlock Shale) your original and typical section and your order of superposition are false. My order of superposition was not false in any essential part..... You have acted contemptuously, unjustly and falsely towards me. I cannot smooth over the matter by the shallow gloss of vulgar courtesy or by abuse of the name of friendship."

It was left to the next generation of geologists to resolve this dispute. Here we meet Charles Lapworth (1842-1920), who started his career as an enthusiastic amateur – a schoolmaster – and ended it as Professor of Geology at Mason College (which became Birmingham University). He proposed that the disputed strata should be placed in a period of their own, based on their distinctive fauna of graptolites. These fossils resemble floating razor shells or tuning forks. Their white or grey remains can be found on the bedding planes of ancient shales. Lapworth called the new period the Ordovician, after the Welsh tribe, the Ordovices (who gave such trouble to the Romans that they were slaughtered *en masse*.) His new period was eventually accepted worldwide.

\*\*\*\*\*\*\*\*\*\*\*\*\*\*\*\*\*\*\*\*\*\*\*\*\*\*\*\*\*\*\*\*\*\*\*\*\*\*\*\*\*\*\*\*\*\*\*

One famous episode in palaeontology dates from the 1870s. It centres upon the Morrison Formation, a sequence of late Jurassic sedimentary rocks that is found in the western United States. This has proved to be the most fertile source of dinosaur fossils in North America. Most of the fossils occur in the siltstone and sandstone beds which form the relics of the rivers and floodplains of the Jurassic. It is centered in Wyoming and Colorado, with outcrops in many other western states and in Canada.

Thousands of dinosaur fossils have been recovered from this area, including the theropod (upright carnivore) *Allosaurus*, and at least two species of *Stegosaurus*, first described by Othniel Marsh. Sauropods (heavy grazers) found here include *Diplodocus, Camarasaurus* (the most commonly found sauropod), *Brachiosaurus* and *Apatosaurus* (formerly known as *Brontosaurus*). The very diversity of the sauropods has raised some questions about how they could all co-exist. While their body shapes are very similar (long neck, long tail, huge elephant-like body), they are assumed to have had different feeding strategies, in order for all of them to have existed in the same time frame and similar environment.

When news of the discoveries in the Morrison Formation began to filter back east in the United States during the 1870s, it sparked of a period of intense rivalry known as the "Bone Wars". This involved two men, Edward Cope (1840-1897) of the Academy of Natural Sciences in Philadelphia, and Charles Othniel Marsh (1831-1899) of the Peabody Museum of Natural History at Yale. Marsh had a rich uncle, George Peabody, who had made his money in banking. He not only paid for Marsh's education, but for the Peabody Museum and the job of running it, which of course went to Marsh. Cope was also well-funded. All sorts of skullduggery was practised in the competition, including smashing up bones once a find had been boxed up in a case.

The great aim of American palaeontology at this time was simply the naming of new species – for the name of the finder is generally attached to the species name and stays with it forever. There was much less interest in the hard grind of palaeontology, which involves identifying the fossils which can be used to correlate different beds in locations spread over a whole country or indeed the whole world. Dinosaur bones are not much use for that (but they are VERY good as museum exhibits!)

## Chapter 8 – Life Sciences in the Nineteenth Century

The first figure of note in Victorian life sciences, a contemporary of Charles Darwin, was Sir Richard Owen (1804-92). This man was the driving force behind the establishment of the Natural History Museum in South Kensington, still the top choice for any child or adult on a visit to London and famously home to *Diplodocus* and many other dinosaurs. It was Owen who coined the very term dinosaur ("terrible lizard"). He also made a remarkable discovery about the anatomy of tetrapods – creatures with four limbs; that all their limbs follow the SAME blueprint. In the case of the human arm, this is one large upper bone (the humerus) which then articulates to two bones (ulna and radius), which then attach to a series of small bones at the wrist (carpus), before spreading out into five digits or fingers, each made up of several separate bones. The human leg follows the same pattern. Owen, however, noted that this system of bones was essentially the same in all tetrapods, be they frogs, birds, pterosaurs, bats, lizards, seals, theropod dinosaurs (upright carnivores) or even humpback whales. He published his findings in a book called *On the Nature of Limbs* in 1849, ten years before *Origin of the Species* appeared. Darwin, far from being ahead of his time, was only just in time with his theory. The common architecture of limbs points to one, now very obvious conclusion – that all tetrapods evolved from a single common ancestor, now thought to be a lungfish which clambered out of the water in the Devonian period nearly 400 million years ago.

\*\*\*\*\*\*\*\*\*\*\*\*\*\*\*\*\*\*\*\*\*\*\*\*\*\*\*\*\*\*\*\*\*\*\*\*\*\*

The intellectual successor of the geologist Charles Lyell was Charles Darwin (1809-1882), one of the most influential thinkers in the history of the planet. He was born at the family home called the Mount in Shrewsbury, county town of Shropshire, the son of a prosperous doctor. As noted, his grandfathers were Erasmus Darwin and Josiah Wedgwood, the wealthy potter. Erasmus Darwin was a considerable naturalist in his own right, and had produced a well-known book called

*Zoonamia* (1794-6). This work firmly espoused the idea of the evolution of species, and suggested that all warm-blooded animals sprang from one original "filament". It also suggested that the fittest species are the ones which survive to propagate themselves. However this elder Darwin failed to hit upon the right method of evolution, believing that improvements made during the life of a single creature could be passed on to the next generation, which they cannot. For example a giraffe, by stretching to eat leaves, would develop a longer neck, and pass on this characteristic to the next generation. We now know that characteristics acquired in this way cannot be passed on. The idea may seem like common sense, but once again, common sense is wrong.

In the next generation, the Frenchman Jean-Baptiste Lamarck published another mammoth work entitled *Natural History of Invertebrates*, the first volume of which appeared in 1815. "Invertebrates" is a term he himself invented for creatures without backbones – in fact a very diverse group including all worms, shellfish and insects. Like Erasmus Darwin, Lamarck was a firm believer in evolution, by now an accepted fact amongst many scientists, but he too believed that changes which took place in the lifetime of a species could be passed on to the next generation, called the Lamarckian system after him.

So Charles Darwin himself had an important background in evolution, but his first real career was as a geologist. As a young man he appeared lackadaisical, causing his widowed father to complain "You care for nothing but shooting, dogs, and rat-catching, and you will be a disgrace to yourself and all your family". He began training as a doctor, but one sight of child surgery without anesthetics sent him running for cover, and in fact gave him nightmares for years afterwards. At the age of 22, after spending time at Edinburgh University and Christ's College, Cambridge, from which he emerged with a degree in divinity, the only route forward seemed to be the one that led to a country parsonage.

At that moment Darwin was suddenly confronted with an amazing offer – to spend a number of years abroad on the survey ship *HMS Beagle*. Its Captain, Robert Fitzroy, wanted a companion from his own level of society, as the conventions of the day would not allow him even to share a meal with other ranks. His predecessor as captain had put a bullet through his own head. (Fitzroy DID eventually commit suicide in 1865, as had his uncle, the senior politician Lord Castlereagh.) The ship set sail at the very end of 1831, when Darwin was only 22, and Fitzroy only 26. Darwin took with him the first volume of Lyell's *Principles of Geology*, received the second volume whilst abroad, and read the third when he finally arrived home in October 1836. (This was to be Darwin's only foreign excursion.)

Fitzroy must have been more indulgent than his quarrelsome reputation suggests, because Darwin in fact spent months at a time on land, in South America. Here he saw tectonic activity first-hand, as whilst in Chile he witnessed an earthquake. When the dust had settled, the level of the ground had been raised several feet. As he had also witnessed the rapid erosion of the Andes, he was very much inclined to agree with his friend Charles Lyell! It was perfectly obvious that the processes at work on the Earth now could also explain much of the geological record.

Darwin was deeply impressed by many of the animals he saw in South America, which after over a hundred million years of separation from the rest of the world (we now know that it split from Gondwana during the Cretaceous period) had developed a very distinctive fauna. Creatures such as the gaudy cracid (rhymes with acid) birds are found nowhere else, and had clearly descended from common ancestors. However he didn't think much of the human inhabitants of Chile, regretting the fact that it had been settled by the Spaniards, instead of the English, who in his opinion would have made a much better job of it!

The Beagle's most famous port of call was the Galapagos Islands, which certainly made an impression upon Darwin: "The country was compared to what we might imagine the cultivated parts of the Infernal Regions to be." The islands are purely volcanic and have never been in contact with the mainland. They have only two native mammals, and five reptiles, including the famous giant tortoises. Darwin also counted fourteen species of finch:

6 species of ground-finches, feeding on seeds in the arid coastal regions

6 species of tree-finches, feeding on insects in the moist forest regions

One warbler-like finch eating insects in both the wet and dry regions

One Cocos Island species living off insects in the tropical forest.

These were his comments. The size and shape of the beak varies according to the bird's diet. Of the second group, the woodpecker finch is remarkable because it has evolved the beak but not the long tongue of the woodpecker, so it has to use a twig or a cactus spine to dislodge insects, a rare case of bird technology. The three largest insectivorous finches would not normally be considered separate species, but they live on separate islands without interbreeding. In the million or more years or so of its existence, the isolation of the Galapagos Islands has allowed this speciation of the finches. The fact that there are several separate islands encouraged further speciation – on Cocos, which is only one island, there is only one species of finch.

Arriving back in England already something of a celebrity, Darwin set about converting himself into a naturalist – in fact a barnacles specialist, eventually claiming that "I hate a barnacle as no man ever did before"! He went on to become Britain's first popular scientist, eventually producing nineteen books. The first of these was a best-seller, *Voyages of the Beagle* (1839).

Locating himself at Down House in the village of Down in Kent, Darwin lived the life of a gentleman scientist, producing a large family and enjoying modest fame. He had certainly developed his theory of natural selection by 1842, but knowing its publication would unleash a storm of criticism, he kept it to himself. In 1844 a book was published called *Vestiges of the Natural History of Creation*, by an anonymous author, which had the temerity to suggest what Darwin was also about to suggest, that man had descended from the apes without any need of a divine creator. The author was in fact a Scottish publisher called Robert Chambers, and his book was blasted from every pulpit in the land.

Darwin was pushed into the publication of his great book, knowing that it would upset his wife and her simple religion, by an old-fashioned scientific impetus – a priority claim. In 1858, another British naturalist, Alfred Wallace, produced a paper which amounted to a summary of Darwin's unpublished book, and sent it to Darwin himself for comment. "If Wallace had my manuscript sketch written out in 1842," he exclaimed, "he could not have made a better short abstract."

Wallace (1823-1913) was born in Monmouthshire, then in England, now in Wales. He and his brothers were active in surveying and building, a precarious business in times, then as now, of boom and bust. In 1847 Wallace took the plunge and disappeared in the Amazonian forests of Brazil with the idea of collecting new and strange specimens of plants for aristocratic collectors in Europe. He spent four years here – an experience not dissimilar to Darwin's, though with no comfortable cabin back on board ship. He set off once again in 1854, this time to the Malay archipelago, a much less well-known area scarcely explored by European botanists. Whilst still in the area he produced a paper on the subject of evolution by a process of natural selection, and sent it to Darwin, who in turn sent it to Charles Lyell. The upshot was that a joint Wallace and Darwin paper, announcing the new theory, was read to the Linnaean Society in 1858. This produced little comment; but Darwin had now been pushed into producing his great book, *Origin of the Species*, which emerged the following year.

Far from being insulted, Wallace was delighted to be linked to the great man, and his name lives on in botanical circles to this day. There is a second reason for this. He observed a zone of limited mixing of the Australian/New Guinean and Asian biota in the Malay archipelago, but noted that the line between the two is still quite distinct in these island territories. He published a paper on the biogeography of the area in 1859. The line, ever since known as the Wallace Line, is one of the most famous in biology. It passes between the islands of Sulawesi (Celebes) and Borneo, with New Guinea and Australia on the south-eastern side of it, and Java, Sumatra, Malaysia and the Philippines on the north-western side.

*Origin of the Species* did not create the idea of evolution, which was already well established, but a mechanism for it, natural selection, or more specifically the competition between different members *of the same species* for survival – "survival of the fittest" (not Darwin's own words). Darwin's idea was that minute (genetic) differences between the offspring of any species, as brought about by mutations, could confer an advantage in the highly competitive environments in which most creatures live. In that case the mutation would survive and prosper in future generations. If such a mutation had a deleterious effect then its bearers would not survive to reproduce successfully. As tiny increments of mutations would take many generations to become established, it was clearly going to take a vast amount of time to produce all the creatures on the Earth, let alone all the extinct creatures identifiable from the fossil record. Yet, fundamentally, it was such a simple idea.

Natural selection not only affects the ability of an animal to feed itself and survive predation. It needs to reproduce as well, and here Darwin noted some of the truly remarkable features produced by sexual selection which would otherwise seem a distinct disadvantage, most famously the tail of the peacock. Nobody really knows why the dowdy peahens like mates with such splendid feathers, but clearly, they do, even if it means that their mates can barely fly, and thus find it much more difficult to escape from hungry leopards. Again, the males of most deer species have quite ridiculous headgear, with which they batter each other – sometimes to death – in the mating season; but without it, they do not reproduce.

Darwin had been influenced by Thomas Malthus who in 1798 published *An Essay on the Principle of Population*. In this he stated that human populations increase exponentially, and could double with every generation, whereas food production can only increase at lower, arithmetic rates. Therefore sooner or later the population would have to be reduced by famine, warfare or disease. Darwin saw that the same pressures of food supplies and predation applied to animal species. If a stable numbers of any animal species were to be maintained, many would have to die without reproducing first, as birth rates generally greatly exceed the rates needed merely to maintain the same population.

In fact Darwin had already observed the amazing changes brought about by the selective breeding of pigeons and other domesticated species, and he saw that natural selection operated in the same way. In other words Darwin was looking at what we would call genetic changes brought about by selective breeding, either natural or artificial. Small changes occurring naturally between one generation and its offspring could in the end produce whole new species, as had clearly happened with the finches of Galapagos. Nowhere did Darwin suggest that these changes came about because of any pattern of behavior of the parents in the Lamarckian manner. In fact most of these changes will be deleterious, and will not produce a fitter creature in the next generation. Indeed humans themselves have characteristic genetic mutations which do not produce better people. Instead they produce people with Down's Syndrome, hunchbacks and dwarves; and people with spin bifida, Hutchinson's chorea and other genetic diseases.

After publication, Darwin had to face the music, tormented as he was by his own theory – "like confessing to a murder". The first edition of his new book, 1,250 copies, sold out on the first day, and soon the knives were out. However Darwin did to some extent leave it to others, notably Thomas Huxley, to state his case in public. In a famous debate held in the Oxford Zoological Museum on 30 June, 1860, Huxley took on Samuel Wilberforce, the Bishop of Oxford. After Wilberforce demanded of Huxley whether his ancestry from the apes came by way of his grandmother or grandfather, the meeting collapsed in tumult, both sides claiming victory.

If evolution happened by a series of small steps, the opponents argued, then where are the intermediate species? In fact the geological record IS short of intermediate species, simply because the preservation of fossils is such a haphazard process. Bats, for example, first appear fully formed, with no hint as to how they got to fly. The most famous intermediate species, *Archaeopteryx*, part bird and part reptile, was actually discovered in Bavaria in 1861, and might have helped Huxley in the debate a year sooner. Even today, this debate still goes on. Some palaeontologists see evolution not as a steady process, but coming in sudden bursts – punctuated by periods when nothing much seems to change.

Also, even Darwin conceded that his process of natural selection required enormous amounts of time – deep time – to work. As late as 1900, physicists such as Lord Kelvin were prepared to give the Earth an entire history of no more than 100 million years, yet it was known that the large thicknesses of older strata (the Precambrian) contained no sign of life, so the timescale simply looked too short. Darwin found himself very much on the defensive, despite the rather obvious similarities between people and chimpanzees!

Darwin went on to produce a total of nineteen books of popular science, as well as further editions of *Origin of the Species*. One of his books was *The Descent of Man*, which was far more specific about man's ancestry amongst the apes than *Origin of the Species* had been. However he struggled with the question of how inheritance took place – in fact he got it completely wrong. He thought that an essence was somehow passed from every cell in the body into the next generation, a process known as pangenesis, where a kind of averaging took place of the material received from either sex. He gradually shifted towards a Lamarckian point of view that acquired characteristics could be passed on by inheritance. Had he been a statistician he would have realised that the blending of parental characteristics which he proposed would have produced a more or less uniform population within a few generations – the exact opposite of what is required for natural selection to work.

Natural selection as described by Darwin is an amazingly powerful process, but it is not random – far from it. Instead it is distinctly nonrandom and purposeful. For example it has managed to create three anteaters in different continents – the echidna of Australia, the pangolin of Africa and the giant anteater of South America, which share a large number of features. These include a long and hairless snout, a long, sticky and versatile tongue, large salivary glands, a rugged stomach (swallowing termites involves swallowing a lot of sand), vestigial teeth, and digging, scraping claws. Yet the creatures are not remotely related. The echidna is a primitive creature, one of only two mammals which lays eggs (the other is the duck-billed platypus). In geological time, natural convergence, as this process is called, has given us the swordfish (a true fish), the dolphin (a mammal) and the ichthyosaur (a reptile of the dinosaur age), all of which look quite similar to the untrained eye. In the plant kingdom, there are many convergences, including the euphorbia group in Africa, which closely resemble the cactus group of the Americas, to which they are quite unrelated.

The theory of natural selection has been central to biology ever since Darwin's time, and some remarkable insights into the way it works have been noted. It does not plan ahead, it uses whatever material comes to hand, and it sometimes seems to operate in almost a slapdash way. In Richard Dawkins' ringing phrase, it is a blind watchmaker. One of the best example of this is its attempts to do something about the scourge of malaria. In West Africa up to two-thirds of all children carry at least one copy of the gene which produces the red blood pigment, haemoglobin, which has the mutation known as sickle cell. When a cell in the body is attacked by the malaria parasite, this haemoglobin mutation causes the cell to collapse. A child with a single copy of the gene gains a protection against severe symptoms which is 90% effective. Now comes the bad news. If the child has two copies of the gene it is likely to develop a fatal disease called sickle cell anaemia.

Other minor haemoglobin changes are noted in the inhabitants of India and the Middle East, which are similar in effect to sickle-cell – they cause an infected cell to commit suicide. However, in some Mediterranean cases, notably in Cyprus and Italy, whole sections of the haemoglobin gene are deleted, again giving rise to serious illnesses known as thalassaemias ("sea blood"). Children inheriting one copy of the mutated haemoglobin gene obtain relief from malaria, but those with two copies are unable to make red blood cells properly. In cases such as sickle-cell disease and thalassaemia, natural selection has picked up the nearest mutation to hand in a desperate situation, with drastic consequences in some cases.

*Origin of the Species* was just as influential for botanists as for zoologists, because of the bright light it shone on the hitherto baffling business of the relatedness of species. It was now clear that the closeness of species all depended on how far back in time they shared a common ancestor. This in turn explained something about plant hybridization. The very first artificial hybrid plant, a cross between a sweet pea and a carnation, had been created as long ago as 1717 by Thomas Fairchild. He produced a real plant and flower, but no seeds – the common ancestor of the parent plants was much too distant. This was now understood for the first time – and hybrid plants were to go on to feed the world, so this was important.

\*\*\*\*\*\*\*\*\*\*\*\*\*\*\*\*\*\*\*\*\*\*\*\*\*\*\*\*\*\*\*\*\*\*\*\*\*\*\*\*\*\*

Just at the time when Darwin was publishing his famous *Origin of the Species*, another man a long way away was producing results which, had he only known about them, Darwin would have found very useful indeed, especially for some of his later work which dealt with this very subject – the natural means of inheritance. This man was an Austrian, Gregor Mendel (1822-1884), whose misfortune it was to achieve nothing but scientific obscurity in his own lifetime, and great fame after it was over, for he is the undisputed father of a whole new and very important science, genetics.

Mendel came from a poor farming family in Moravia in what is now the Czech Republic, but he made his way through the gymnasium and spent two years at a scientific institute in Olmütz. In 1843 he became an Augustinian monk in the monastery at Brünn (modern Brno), where the abbot was gathering a coterie of bright young men in different disciplines. In fact it was a mini-university with a library of 20,000 books. For Mendel, this was a way of furthering his education, and in fact he was to spend two years at the University of Vienna from 1849, where he studied, amongst other things, statistics, plant physiology and the atomic theory of chemistry. Hence he was NOT a simple monk tilling a garden, as he is sometimes portrayed.

Mendel then spent seven years from 1856 on agricultural trials to try to get to the bottom of the principle of heredity. Remember that Darwin thought that heredity worked by blending, or averaging characteristics. Mendel was to show that there is no averaging – all inheritance comes from one parent or the other, unchanged in almost all, but not quite all cases.

This was no small-scale enterprise. Mendel cultured 28,000 plants and subjected nearly half of them to careful examination. He and his two full-time assistants had to take the utmost care in pollinating his plants with known parents, to avoid accidental pollination. In fact he and his assistants had to pollinate the plants by hand, and keep records of what they had done.

His most famous results were achieved with pea plants, which exist in true-breeding lines where all the individuals of one variety look the same, but different from the next variety. (In fact to ensure the purity of his varieties, Mendel spent his first two years preparing seven different varieties of pea to make sure that they did breed true.) The main differences are in skin colour – yellow and green – and in surface texture, smooth and wrinkly. These are qualities which can be COUNTED, rather than measured. (The other useful quality of the crop was that it could be eaten after the experiments!) The plants can be cross-bred, or otherwise self-pollinated, so that an individual plant can be made to fertilize itself. This is a crucial point, because Mendel obtained the most startling results by cross-breeding a yellow variety with a green variety, and then self-pollinating the next generation.

In statistical terms, he found that if he started with four pea plants, two yellow and two green, and bred them together, the next generation would yield four pea plants all with yellow peas. Self-fertilizing these by hand, the generation after that produced three plants bearing yellow peas, and one bearing green peas! What was going on here? He decided that the reason must be that each pea plant inherited BOTH instructions for colour. He called these instructions factors, and we call them genes. However, one of these genes must be dominant, and the other recessive, so if both are present, then only the dominant will actually be visible in the pea colour, and in this case the dominant colour is yellow. So mixing greens and yellows the first time round produced nothing but yellow peas, but EACH OF THESE actually carried two genes, one each of yellow and green. When self-fertilised, on average, there resulted one plant with two yellow genes, two with a yellow and a green, and a fourth with two green genes. This same pattern was repeated with skin texture (smooth and wrinkly), but it did not follow the skin colour in the same plants. Mendel concluded that there must be individual factors or genes to control different physical characteristic of the peas.

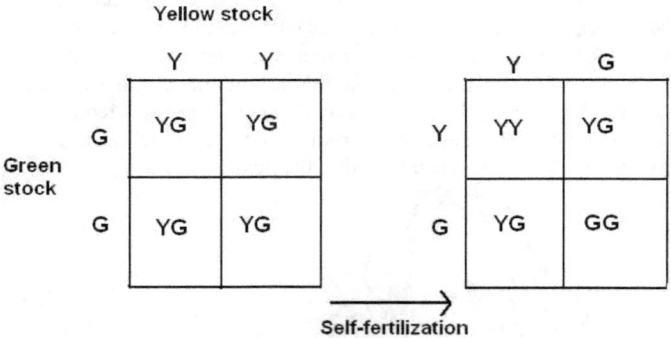

The inheritance of pea colour according to Mendel

Mendel's results were read to the Natural History Society in Brünn in 1865 and published in their *Transactions*, but they failed to attract much attention. Mendel, 42 at that time, was elected abbot in 1868 and gave up his plant breeding program. He did circulate his paper to the people who mattered in botany, and it was subsequently cited in some important papers and achieved a reference in the *Encyclopaedia Britannica*, but otherwise it was forgotten for forty years – truly an idea ahead of its time.

## Chapter 9 – Victorian Science

Davy's assistant Michael Faraday (1791-1867) achieved success on his own account, of course, and rather late in life for a scientist – some of his best work was done when he was over 40 years old. One reason for this was that he had to work his way up from a rather low rung on the ladder of society. His father was a blacksmith, originally from Westmorland in the Lake District of NW England, though the family had moved to London by the time Michael was born. They were members of a Protestant sect, the Sandemanians, and when Faraday eventually married, it was into this community. In adolescence the young man found himself apprenticed to a bookbinder, an apparently kindly gentleman by the name of George Riebeau. In 1810, the year his father died, Faraday became a member of the City Philosophy Society, where he attended lectures and scrupulously wrote up the notes. In 1813 he hit pay dirt – he managed to obtain an appointment as assistant to Humphrey Davy, partly on the basis of a recommendation from a Mr Dance, who had been impressed by the young man.

As noted, in 1813 Faraday set off with Davy and his new wife on a tour of western Europe which took 18 months, and required a special passport as England and France were at war. The French were only too happy to welcome the famous scientist. Davy's valet lost his nerve at the last moment and refused to travel to Napoleonic France, which meant that Faraday had to fill in for him, not something he enjoyed. Nevertheless, despite being treated as a servant by Davy's wife, he made the most of the experience. By the time he came home he had met most of the famous scientists of Europe and had learned to read French and Italian.

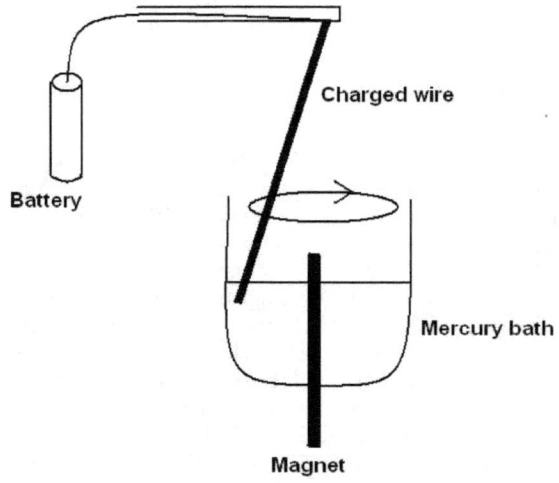

Faraday's first experiment to produce motion by electromagnetism

In 1820 a Dane called Hans Christian Oersted made the first breakthrough in the field of electromagnetism. He noted than when a magnetic compass needle is held over a wire carrying an electric current, it is deflected at right-angles across the wire. This was utterly unexpected as it suggested that the electric current in the wire was creating a circular magnetic field around it. All of scientific Europe pricked up its ears. A British scientist called William Wollaston, the man who discovered palladium and rhodium, arrived at the Royal Institution to try out some experiments of his own on this new force. These did not work, but he discussed them with Faraday afterwards. Faraday then constructed his own experiments and these did work. The first of them involved a fixed magnet immersed in a bath of mercury. A wire was also suspended into the mercury, and this wire was connected to a battery. The result was that the wire circled around the magnet. Faraday had created the first electrical motor – that is, he had used an electric current to make something MOVE. This was the first new form of motion that did not involve either animal, water, wind or steam power. It was only a simple device, but the potential was immediately apparent. He also conducted a second, similar experiment in which the magnet revolved around a fixed wire through which a current ran. In his excitement, Faraday rushed to publish his results, without acknowledging Wollaston. This strained his relationship with Davy, but the other fellows of the Royal Society were certainly impressed. Hundreds if not thousands of experimenters all over the western world were trying to achieve results like these.

In 1831 Faraday discovered electromagnetic induction, whereby a current in one wire creates a momentary current in another. Insulated wires are wrapped around either side of an iron ring (like a pair of ears) to obtain this effect. The wire receiving the electric charge acts like a magnet to induce a current in the other wire, but the effect is only momentary, and appears only when the current is switched on or off.

By extention, he also established that if he moved a magnet into a coil of wire, then an electric current was induced in the wire. So just as a current (moving electricity) can induce magnetism, so a moving magnet can induce electricity. This explained why nobody had ever been able to induce an electric current by using static magnets. Faraday had now invented the prototype electric generator or dynamo, which uses the relative motion of coils of wire and magnets to generate electricity. A dynamo is an electromagnetic rotary device in which magnets spin around a coil of wire. Many large modern generators work the other way round, with a wire coil spinning round inside a set of magnets, so generating electricity. The turbines which drive the coil are in turn driven by water power, or by any other fuel.

Faraday became quite heavily involved with electrolysis, and in 1834 he coined the term ION, meaning something which GOES. He knew, like Berzelius, that the copper in a solution of copper sulphate would find its way to the cathode or negatively charged electrode. He termed such substances CATIONS, or positively charged ions, and those accumulating at the anode, ANIONS. However although he realised they must be electrically charged, chemistry had not advanced far enough to permit him to know what exactly was going on. We can now define an ion as an element or molecule which has an unbalanced number of electrons. For example, hydrogen invariably has one positive proton at its centre. It normally also has one negative electron and so is electrically neutral. If it should LOSE this electron, it becomes positively charged (a cation). If instead it GAINS an electron, it becomes an anion.

Later in his career, Faraday proposed that electromagnetic forces extend into space from conducting bodies, including the Earth and the Sun. He claimed that magnetic, electrical and gravitational forces penetrate the empty space of the universe, and although they act at a distance, they need time to cross space. In so doing he meant to "dismiss the ether", the substance held to fill the void of space since the time of Aristotle. He was right, but this was not apparent at the time!

Faraday eventually died, childless, in 1867, in the grace-and-favour (i.e. free) apartment which had been granted to him at Hampton Court in 1848. In his lifetime he had refused many honours, including a knighthood – such baubles were meaningless to him. His picture appeared on £20 notes issued by the Bank of England from 1991, and he certainly looks like a clever man. An SI unit, the farad, is named after him. This is the measure of capacitance, or the charge in coulombs which a capacitor will accept. Widely acknowledged as one of Britain's greatest scientists, his inventions mark the start of the modern world based on electrical machinery.

*****************************************

The most famous botanist of the Victorian era is the German Julius von Sachs (1832-97), who carried out his most important work at the University of Würzburg from 1868. New, more powerful microscopes than had hitherto been available enabled him to identify what were later called chloroplasts as the source of chlorophyll in green leaves. Experimenting with leaves exposed to the sun, not exposed, and partially exposed, then dyed with iodine, he realised that starch, the first product of photosynthesis, is only made where the leaf, or part of the leaf, has been exposed to sunlight (sugar is then made from the starch). The new microscopes also enabled him to identify minute pores on the bottom of leaves, which he noted could be opened and closed to admit or in certain circumstances to expel carbon dioxide. The understanding of the process of photosynthesis, which had begun with Van Helmond and continued with Ingenhousz, was nearing completion, though the main breakthrough on the route taken by carbon in this process would have to wait until the middle of the following century. Al least at this point, it was clear that the main inputs were carbon dioxide, water and sunlight, and the main outputs were carbon, used by the plant itself, and oxygen, expelled as a waste product – not, however, a waste for the animal kingdom, most of which could not exist without it!

*********************************************

In the middle part of the nineteenth century, chemistry finally came to terms with the idea of atoms and the fact that electrical charges bind them together into molecules. In 1852 the English chemist Edward Frankland gave clear expression to the idea of valency, or electrical equivalence, by which bonds are created between elements. Hydrogen has a valency of one, which means that it can bond with one other element. Oxygen has a valency of two. So two hydrogen atoms can link with one oxygen atom, in the form H-O-H. Nitrogen has a valency of three, so it links with three hydrogen atoms to form ammonia, $NH_3$. Bonds can also form between atoms of the same element, as with molecular oxygen $O_2$, or O=O. Carbon, which lies at the heart of all organic chemicals, has a valency of four, and this gives it great flexibility. The simplest organic compound is methane, $CH_4$, where one carbon atom links with four hydrogen atoms.

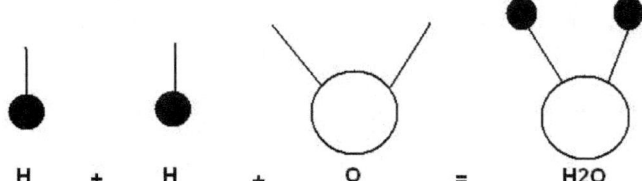

Bivalent oxygen combines with single-valency hydrogen to form a water molecule

Carbon atoms can also link up in rings, commonly in a ring of six atoms forming a hexagon, with bonds forming outwards to other molecules, eventually forming chains of very complex organic molecules. These concepts were developed by the German chemist Friedrich Kekule (1829-1896).

In the year 1860 the first international conference of chemists was held in Karlsruhe in Germany. As the meeting was closing, the Italian chemist Stanislao Cannizzaro handed out a pamphlet which he had written on atomic weights, following up on fiery, inspirational speeches he had given on the same subject at the conference. It was no less than a skeleton version of what would become the periodic table of elements, in which all the elements are linked in groups according to their atomic weights and properties. For many of the chemists present, including the Russian Dmitri Mendeleyev, this was the moment when the scales fell from their eyes – it was the start of order emerging from the previous chaos.

During the course of the 1860s, no less than four attempts were made – from England, France, Germany and Russia – to group the known elements together based on their properties and atomic weights. It was clear that there was some sort of relationship between the members of each group of elements, which were always separated by multiples of eight times the atomic weight of hydrogen. This led the English industrial chemist John Newlands to propose a "law of octaves" – the eighth element from any starting point is a kind of repetition of the first, exactly as in a musical octave. He noted the similarity between the alkali metals sodium and potassium, between magnesium and calcium, and the halogens iodine, bromine and chlorine (plus fluorine) as previously noted by Döbereiner. The chemists seemed to be getting tantalizingly near to a glimpse of the very blueprint of the universe.

However there was plenty of opposition from the chemical establishment in all western countries, and Newlands' audiences were inclined to ask him if he could get his elements to play a little tune. The man who won the race for priority came from far, far away – he was Dmitri Mendeleyev (1834-1907), the very epitome of the mad professor, brilliant, highly irascible and extremely hairy (said to have trimmed his beard but once a year), and he came from St Petersburg.

Mendeleyev's path to his professorship had been far from easy. He was born at Tobolsk, in Siberia, youngest of fourteen children. His father, a local headmaster, first went blind, and then died in 1847. His mother, an indomitable woman by the name of Marya, set up a glass factory to support the family, but this was destroyed in a fire in 1848. Undeterred, she set off to St Petersburg, determined to get the best possible education for her youngest child. Mendeleyev managed to enroll as a student teacher in 1850, whereupon Marya also expired. Nevertheless he worked his way up the ladder and was a professor by 1864.

Mendeleyev's first attempt at a table of the elements had them arranged in rows of eight across the page, with those with similar properties arranged in columns underneath. The lightest element was in the top-left corner, and the heaviest at the bottom-right. However, there were some discrepancies. For example, in order to make them fit in the right place in the table, he had to swap the positions of tellurium (weight 127.6) and iodine (126.9), as iodine clearly belonged under bromine, and not tellurium. A second problem was that there were gaps in the table, but as new elements were being discovered all the time, this did not bother Mendeleyev. He predicted that new elements would be found to fill these gaps, and was also able to predict what some of their properties would be. It was this boldness – first in switching some of the elements, secondly in assuming there would be new elements – which led Mendeleyev to be regarded as the father of the periodic table. He published his first version of it in a paper entitled *On the Relation of the Properties to the Atomic Weights of the Elements* in 1869.

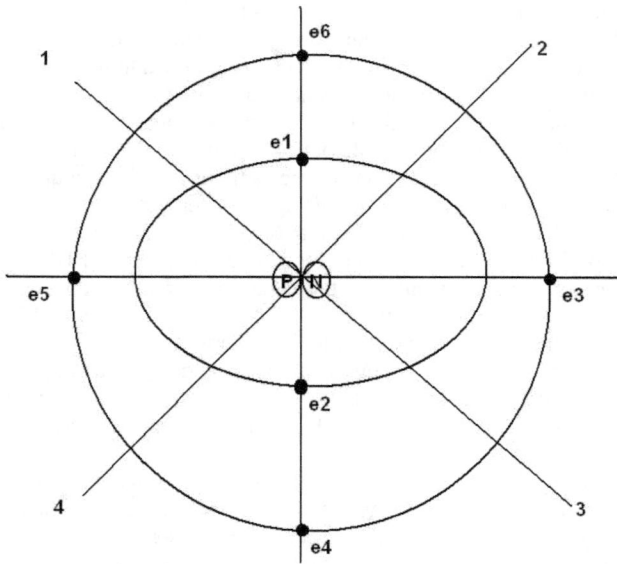

The cause of Mendeleyev's confusion. The above represents a carbon atom. It has an atomic number of 6 because it has 6 protons in its nucleus. Corresponding to this, it also has 6 electrons, two in the inner ring and four in the outer ring, leaving four vacant slots (valency of 4) left over to combine with other elements or molecules. The commonest isotope also has 6 neutrons in the nucleus, giving an atomic weight of 12, but this can vary as there are other isotopes with 7 (carbon-13) or 8 (carbon-14) neutrons. The existence of isotopes was unknown in 1869.

The reason for the apparent discrepancies in the table is now known. The place of an element in the table depends on its atomic number, which corresponds exactly to the number of protons in its nucleus and is always a whole number. The atomic weight also depends on the number of neutrons, and this can be variable between different isotopes.

By 1871 Mendeleyev had included all 63 known elements in his table but there still remained three gaps. As he had foreseen, these were filled in during the next 15 years with the discovery of gallium in 1875, scandium in 1879 and germanium in 1886. In the case of germanium, a dark grey metallic element with properties between those of silicon and tin, Mendeleyev had not only been able to predict this physical description but also its atomic weight, its specific gravity, and the specific gravities of its oxide and of its compound with chlorine. Gallium turned out to be a metal with a very low melting point – it melts in the hand.

As there were other competing classifications, Mendeleyev's periodic table was slow to gain acceptance, but the filling of the gaps gave it credibility. The first serious threat to the scheme came in 1895 when the Scottish chemist William Ramsay discovered the first of the noble gases, argon. The noble gases do not react with any other elements and they did not fit anywhere into the periodic table. Mendeleyev, at first sceptical, solved this problem by simply adding a whole new column for them.

The periodic table of elements has been revised many times since Mendeleyev's time, but it has retained its basic shape. Now based on atomic numbers rather than weights, it runs from hydrogen at the top left, atomic number 1, to uranium, the heaviest naturally-occurring element, atomic number 92, plus other heavier "elements" which have been created in laboratories. It has proved itself to be a most elegant, useful and logical chart.

Mendeleyev himself got into a scrape with the authorities in Russia in 1891 when he was retired from his professorship at the age of only 57 for taking the side of the students in a protest about their conditions. He was rehabilitated and given another job three years later, in fact as the controller of the Bureau of Weights and Measures, a position not dissimilar to the one Newton had once held at the Royal Mint.

\*\*\*\*\*\*\*\*\*\*\*\*\*\*\*\*\*\*\*\*\*\*\*\*\*\*\*\*\*\*\*\*\*\*\*\*\*\*\*\*\*\*

One of Dalton's pupils in later life was James Joule (1818-1889), who came from Salford, now part of Manchester. Joule's father was a member of the "beerage", the rich brewers of the nineteenth century. Joule himself ran the brewery at one point, but the young man was always very interested in discovering what exactly went on inside machines, and he is one of the founders of the study of heat and motion, or thermodynamics as it came to be called.

Joule made many attempts to establish a common unit of "work" – the amount of work, or heat, required to raise one pound in weight by one foot – a "foot-pound". Another measure was the amount of heat required to raise the temperature of one pound of water by one degree Fahrenheit. One of his experiments involved the use of a weight (pulling downwards under the force of gravity) to turn a paddle wheel which was inserted into an insulated barrel of water. He then measured the increase in the temperature of the water which had ultimately been brought about by the work done by the descending weight. All this depended on measurements of tiny fractions of one degree Fahrenheit, and the establishment as represented by the Royal Society in London was dismissive of this provincial dilettante. (He eventually established that it took 772 foot-pounds of work to raise the temperature of one pound of water from 60 to 61 degrees Fahrenheit.)

Overall, Joule's researches showed that there is an equivalence between work and heat – that work can be converted into heat, and heat into work (as in a steam engine). Joule challenged the caloric model of heat and showed that energy is conserved as it is passed from one form to another.

One of Joule's supporters was the young William Thomson, with whom he found common ground at a meeting of the British Association for the Advancement of Science in Oxford in 1847. Thomson was to go on to formulate the laws of thermodynamics – the first of which states that heat is the equivalent of work. The Royal Society eventually relented, and elected Joule a fellow in 1850; then awarded him its prestigious Copley Medal in 1870 (years after his best work, which had been concluded by the time he was 30). The SI unit of work is named after him – the joule.

William Thomson (1824-1907) was born in Belfast, where his father was a university professor. Thomson himself became the professor of natural philosophy (i.e. physics) at the University of Glasgow, where his father had moved, in 1846, when he was only 22 years old (he had been a noted child prodigy and had attended the same university at the age of ten). As he held the appointment until his retirement at the age of 75, he had plenty of time to get things done. During his lifetime he was very active in applied science, and was the man behind the first successful transatlantic telegraph cable in 1866 (two previous attempts, where he had not been involved, had failed). In 1892 he was created Baron Kelvin, taking this name from the small river which runs through the site of the University of Glasgow. He is often referred to as Lord Kelvin in scientific circles, to distinguish him from the physicist J J Thomson, to whom he was not related.

Thomson carried on the work of Joule in the area of thermodynamics. As early as 1848 he established an absolute scale of temperature. This is based on the idea that a certain change in temperature corresponds to a certain amount of work. It carries with it the idea that there is a minimum temperature (-273.15 degrees Celsius) at which no more work can be done, because there is no more heat to be extracted from a system or environment. The temperature scale based on degrees Celsius, but starting at -273 instead of zero, is today called the Kelvin scale. (The third law of thermodynamics states that temperatures can never in fact reach zero degrees Kelvin, as there will always be some residual warmth.)

Thomson established the first and second laws of thermodynamics. The first states that heat is the equivalent of work and that energy is always conserved – it cannot be created or destroyed. However, the principle of the conservation of energy is misleading, because not all energy is the same. The concentrated energy of the blazing Sun sustains a myriad of life forms on Earth, but it is eventually radiated back into space as low-grade heat energy which can sustain very little. The energy has not disappeared, but has become so dissipated that it can no longer be used.

The second law is much more important, and states that all closed systems (such as a swinging pendulum) will always run out of energy eventually. The practical effect of this law is to cause a hot substance in a cool place to lose heat to its environment until it is in heat balance with it, rather than to get any hotter. The law also states that systems become more disorganized without an external input of energy – the universe is wearing out. This has proved a fundamental touchstone of physics. If a new theory is promoted which does not fit within this law, its originator is invariably told to think again! Systems which become disorganized are said to possess a high degree of entropy, or state of disorganization. This term was invented by the German physicist Rudolf Clausius (1822-88), who also helped to establish the new science of thermodynamics.

One waggish interpretation of the three laws of thermodynamics is (1) you can't win; (2) you can't break even, and (3) you can't get out of the game! (Quoted in Bill Bryson.)

# Chapter 10 – The later Nineteenth Century

The middle years of the nineteenth century saw further developments in spectroscopy. One of the leading lights in this area was the German Robert Bunsen (1811-1899), otherwise famous for guess what? The Bunsen burner, of course, the most familiar and useful piece of laboratory equipment of all, providing the scientist with a ready and easily controllable flame at last. When heated with such a flame, Bunsen found that specific elements give off characteristic colours. Sodium famously glows bright yellow, which is one reason it is now widely used in street lamps. The flame provides a simple test for the presence of an element, but the spectroscope goes much further, at is was found that each element, when hot, produces a characteristic pattern of bright lines within the spectrum (the Frauenhofer lines). It is as if each element has its own natural barcode. When cold the element will show dark lines (absorbing light) in the spectrum at exactly the same wavelengths which produce the bright lines when it is heated. Bunsen and another scientist working with him at Heidelberg, Gustav Kirchhoff (1824-1887) used this information to detect the presence of previously unknown elements, setting off a new wave of discoveries by spectroscopy as had happened by electrolysis fifty years earlier. Noting deep red spectral lines matching those of no known element, they knew they must have found a brand new one, which they called rubidium. Caesium was discovered in a similar way. These are two of the alkali metals, all of which are soft, shiny and highly reactive, and none of which is found uncompounded in nature. The others are sodium and potassium (discovered by Davy), lithium and francium. Bunsen and Kirchhoff also observed the keynote signature of sodium, indicated by two bright yellow lines in the spectrum, in the atmosphere of the Sun. So for the first time, it became possible to find out what the stars are made of! – without the certainly difficult task of actually going there.

(The Russian chemist Mendeleyev turned up in Heidelberg in 1860 and found to his dismay that despite being a brilliant experimenter,

Kirchhoff had the reputation of being the most boring lecturer in Germany – some feat!)

During the solar eclipse of 1868, the French astronomer Pierre Jansen and the English astronomer Norman Lockyer noticed a pattern of lines in the solar spectrum which did not fit any known element – it must be a new element, discovered at a distance of 93 million miles! Lockyer named this substance helium, one of the noble gases (which do not react with any other elements) and which was not finally identified on the Earth itself until 1895.

Spectral lines were crucial in the discovery of the other noble gases. This work was undertaken by the Scottish chemist, William Ramsay (1852-1916), when following up the observation that nitrogen obtained by chemical reactions was invariably slightly lighter than nitrogen obtained from burning off the oxygen in the air. When the nitrogen left from the air experiment was itself reacted with magnesium shavings, a heavier gas was left behind. It had a brand new spectral signature and was given the name argon (in fact it forms 0.93 percent of the common air). Ramsay then followed up an observation from an American chemist to obtain helium from uranium ore (helium gas is the ultimate product of the radioactive decay of uranium). He went on to discover three more noble gases, krypton, xenon and neon; the final piece in the jigsaw, radon, was discovered a few years later. Ramsay and his assistant were said to have watched spellbound when first passing an electric current through a tube of neon gas, observing the resulting blaze of crimson light in fascination.

The first evidence that these really were new elements came purely from their spectral lines. The gases had no observed physical properties and did not react with anything – hence their name, the noble gases (they are also called the inert gases, though other non-elemental gases are also inert). Neon in particular was to quickly find uses, because it glows a bright red when excited electrically, with no need for any filament. It could be sealed into tubes which then needed no moving parts or chemical reactions to glow happily. The neon advertising industry was born.

\*\*\*\*\*\*\*\*\*\*\*\*\*\*\*\*\*\*\*\*\*\*\*\*\*\*\*\*\*\*\*\*\*\*\*\*\*\*\*\*\*\*

James Clerk Maxwell (1831-1879) is generally considered to hold a position only slightly below Newton and Einstein in the world of physics. He was born in Edinburgh and brought up in Galloway, in the south-western corner of Scotland. His father was a minor Scottish aristocrat. His mother, nearly 40 when she bore James, died at 48 – it is a very common occurrence amongst our scientists to lose a parent at an early age (this happened to Newton, for example). Called "Dafty" at school in Edinburgh because of his outrageous rural accent and home-made clothes, Maxwell completed his education at Trinity College, Cambridge, again the alma mater of luminaries such as Newton and John Ray. He worked as an academic at what became Aberdeen University, and then later at King's College, London.

Some of Maxwell's early work was in the area of optics. He was able to show that a combination of monochrome photographs taken through three different colour filters – red, green and blue – gives a full-colour version. This is still the basis of colour photography, colour TV and computer screens, and inkjet printers. It also helps to explain Dalton's problems with colour blindness – he was missing one of the three natural colour receptors in his eyes.

Maxwell then moved on to study electricity, magnetism and fields of force. Defining his own models, he corresponded with William Thomson to get his ideas sorted out. With mounting excitement he noted that his models and calculations showed electrical and magnetic waves travelling through the air, at right angles to one another, but at a speed which matched exactly the known speed of light. Although light had not entered into his researches at all, he rapidly came to the conclusion that light itself is an electromagnetic phenomenon. In 1864 he set out these ideas in a landmark paper, *A Dynamical Theory of the Electromagnetic Field*. The heart of this was a set of four equations, known ever afterwards as the Maxwell equations. In these he succeeded in uniting all previously unrelated experiments, observations and equations in electricity, magnetism and optics into a single theory.

Maxwell's equations contain a constant, $c$, to represent the speed of light. The remarkable thing about this, which nobody appeared to appreciate for a generation – until Einstein came along – is that it IS a constant. There are important implications for time, for in the standard formula:

Time = Distance/ Speed

2 hours = 60 miles/30 mph

…if speed is constant, then time itself must be variable!

Because of illness, Maxwell was obliged to retire from university work whilst still only 35. He settled comfortably into life on his property in Galloway and produced his equivalent of Newton's *Principia*, entitled *Treatise on Electricity and Magnetism* (1873). He was to die young in 1879, at the same age as his mother – 48 – and also of cancer.

Maxwell's equations predicted that there could be other electromagnetic waves, longer than those of light, and so invisible. In 1886 a German Jewish physicist called Heinrich Hertz (1857-1894) set out to find them, building his own antenna, and succeeded the following year. These were radio waves – and like light waves, it was found that they could be reflected, refracted (bent entering water) and diffracted (spread out). The potential of these waves to carry telegraphic information wirelessly was quickly realised. Within a year, Hertz had demonstrated that it was possible both to transmit and receive them.

Hertz did not survive to witness the wonders of radio, dying at the age of 36. What remained of his family fled Germany for England during the Hitler purges of the 1930s. The SI unit for wavelength frequency, one of the commonest and in daily use by radio broadcasters, is named after Hertz (but he had nothing to do with car rentals!). Marconi carried out his first experiments with radio in 1895, the year after the death of Hertz.

Another Maxwell legacy was the investigation of the concept of the ether, the mysterious substance – weightless, invisible – held to fill the void of space since the days of the ancient Greeks. Its presence seemed more than ever required with the discovery that light travels in electromagnetic waves, or vibrations, because it was felt that these vibrations had to occur in something, rather than in a vacuum.

In an article written in the *Encyclopaedia Britannica* in 1878, the year before he died, Maxwell described a possible experiment to test for the existence of the ether. This works on the assumption that the Earth is physically moving through the ether, and that this will slow down a beam of light.

The beam of light should be split into two. One half should be sent to a distant mirror set up in the direction of the motion of the Earth. The other should be sent to another mirror, the same distance away, but at right-angles to the first beam. When the light is received back from the mirrors, it should be tested for an interference pattern as found in Young's double-slit experiment. If the light comes back without interference, this would mean that the speed of light is the same both in the direction of the Earth's movement, and at right angles to it, and so there can be no ether.

The challenge of setting up such an experiment (named an interferometer) to the required standards of precision was taken up by two Americans, Albert Michelson (1852-1931) and Edward Morley (1838-1923) in 1887 in Ohio. Funding had been provided by Alexander Graham Bell, the inventor of the telephone. This experiment, "to test the ether wind", is in fact the most famous in all of physics. It was repeated in different circumstances many times, but always the result was the same – there was no interference in the reflected light beams, and so no ether wind, and no ether. This negative result gave the first hint in exactly two hundred years that there might be exceptions to Newton's laws. It was interpreted in two ways. The first was to say that the experiment had simply failed. The second, taken by Einstein in 1905, was to say that it had succeeded – not so much in showing that there is no ether, but in showing that the speed of light is always the same!

Michelson, the son of poor Jewish immigrants, was awarded a Nobel prize for his work, the first American to be honoured in this way; but not for twenty years, two years after Einstein had announced his special theory of relativity based on the Michelson-Morley result.

\*\*\*\*\*\*\*\*\*\*\*\*\*\*\*\*\*\*\*\*\*\*\*\*\*\*\*\*\*\*\*\*\*\*\*\*

The belief that matter is made entirely of atoms and then molecules was by no means fully established by Dalton in 1808 – in fact it took almost a complete century before this belief became universal in scientific circles. Many scientists, especially in Germany, believed atomic theory to be merely a heuristic device, and useful only as a rule-of-thumb way of approaching physical problems or calculations. Gradually the mists cleared, and one way this happened was by the development of the kinetic theory of gases. This is because this theory assumes that gases are indeed composed of tiny particles – atoms and/or molecules. It concerns the way these tiny particles move (hence "kinetic", meaning resulting from movement) by bouncing off each other and the walls of their container. As the theory began to receive more and more confirmation from a whole range of eminent scientists, so the underlying assumption was also verified.

We have already met the first significant contributor to this field, James Joule. He published a paper in 1848 which estimated the speed at which hydrogen gas molecules move. Knowing the weight of each molecule and the pressure exerted by the gas, he calculated that this must be 6225 feet (1815 m) per second. As oxygen weighs sixteen times as much as hydrogen, and this weight is represented as a square root in the equation, he found that oxygen gas molecules move at a quarter (the square root of sixteen) of the speed of hydrogen, that is 1556 feet (478 m) per second.

The baton was then picked up by Rudolf Clausius, the man who defined entropy, who introduced the idea of the "mean free path", that is the average distance that a molecule of gas travels before it collides with another molecule. It was then James Clerk Maxwell, greatest of all nineteenth-century physicists, who in 1859 calculated that the mean velocity of molecules in the air is 1505 feet (463 m) per second, and the mean free path 1/447,000 of an inch. This means that each molecule collides with another over eight billion times a second. This work led to a greater understanding of the relationship between heat and motion, as the temperature of an object is a function of the mean speed at which its atoms or molecules are moving. In say a bar of iron, as it is heated, the electrons become excited to a higher energy level – they move faster. (This also causes them to emit electromagnetic radiation in the red part of the spectrum.)

Enter the mad German professor – in fact Austrian, one Ludwig Boltzmann (1844-1906), father of statistical mechanics. This uses probability theory to predict the behaviour of large numbers of tiny particles (molecules or atoms) in thermodynamic systems. That is to say, it became evident that it is not possible to predict what happens to an individual molecule when an object is heated. However the mass of molecules can be subjected to statistical analysis. Boltzmann also contributed to the further understanding of the second law of dynamics – in fact he regarded the probability of any ordered system forming spontaneously in nature as zero, thus confirming the idea that ordered systems, without an input of energy, can only decay into complete disorder, or complete entropy.

Maxwell and Willard Gibbs made further contributions to statistical mechanics and the so the kinetic theory of gases. Gibbs (1839-1903) is the first modern American to make a significant contribution to science.

Boltzmann is mainly remembered today for his contribution to the Maxwell-Boltzmann distribution, a set of equations and curves on a graph which represent the expected velocities of particles moving in a gas. (A distribution in a statistical sense is a plot of values on a graph, the best-known of which is the bell curve or normal distribution.) Boltzmann's ideas relied heavily on the atomic theory, and for this reason he had many problems with the publishers of academic papers and other scientists within Germany. He attempted suicide in 1900, then recovered to the extent that he appeared on American campuses at Berkeley and Stanford, where his eccentric behaviour caused comment. He finally hanged himself in 1906, not knowing that the confirmation of atomic theory had in fact been published the year before.

*****************************************

The next generation of scientists after Maxwell had the benefit of an invention which was to turn the world of physics literally inside out. There followed a period when a whole series of breathtaking new breakthroughs were made, in some cases literally lying around waiting to be discovered. These included electrons, X-rays and ultimately radioactivity. And what was this new marvel of technology? A vacuum tube. Michael Faraday had had to make do with technology no better than had been available to Otto von Guericke two centuries previously, but in the 1850s a German glassblower called Heinrich Geissler (1814-1879) finally invented a tube which used mercury to expel virtually all remaining air and so finally to create a real vacuum. He then sealed electrodes into the tube, and pow! – a whole new world opened up for the scientists. Workers in Germany immediately noticed that when an electric current was passed between the electrodes, glowing rays which cast shadows were emitted from the cathode. These were termed cathode rays, but what were they?

Enter the English. It was here that an experimenter called William Crookes (1832-1919), the first of sixteen children of a tailor and businessman, manufactured his own vacuum tube on the Geissler model, and duly produced the mysterious rays. In the late 1870s he noted that when he placed a Maltese Cross in the tube, its pattern was etched out on the wall of the tube. He also placed a tiny paddle wheel in the tube, which turned round when the current was switched on. Therefore the cathode rays must carry momentum – they must have mass – and they behaved like particles. After the discoveries of Maxwell, linking light, electricity and magnetism, there was a tendency to jump to the conclusion that phenomena of this type must be electromagnetic rays of some sort, but these do not have mass. They must be something else – but what?

Further investigations by various scientists showed that the rays were deflected by electric and magnetic fields, and that therefore they must carry an electrical charge themselves. A German called Kaufmann attempted to establish the mass of the particles using different gases, but he always came up with the same mass for different gases. Technically, he was trying to establish a value for $e/m$, where $e$ is the electrical charge, and $m$ is the mass. The ratio between the two did not vary as expected between different gases.

It is now time for another great scientist to enter the field, J J Thomson (1856-1940), the son of an antiquarian bookseller, born and bred in Manchester but from his twenties onwards another student and then staff member of Trinity College, Cambridge. All very confusing for students of the history of science, this is our third person of this name. The first was Benjamin Thompson, the Anglo-American who exposed the myths of the caloric model of heat by drilling out cannon; then there was William Thomson (Lord Kelvin) who established the laws of thermodynamics. Now there is JJ, the man who discovered the electron!

Unlike Kaufmann, JJ Thomson thought from the start that he was dealing with identical particles emitted as a stream from the cathode, so he was hardly surprised when his own calculations always gave the same value for $e/m$, or as he put it, $m/e$. Expressed this way, he noted the very small number he obtained, as compared with the equivalent result for hydrogen (in fact what we now call hydrogen cations, or single protons without electrons). His figures meant that either the mass of the particle was very small, or the electrical charge very large. However there is no doubt that what Thomson had found was nothing to do with electromagnetic light – it was a subatomic particle, and so the first inkling that something smaller than an atom, previously thought indivisible, might exist. When he gave a lecture on this subject at the Royal Institution in 1897, he observed that "The assumption of a state of matter more finely divided than the atom is a somewhat startling one." Some of his audience thought he was joking.

Two years later Thomson succeeded in measuring the electrical charge itself, using electrically charged droplets of water – this man did not throw money around! This enabled him to obtain a value for $m$, which showed that each particle had a mass of one two-thousandth of the mass of a hydrogen atom. In fact it is indeed very tiny, but (in the case of hydrogen) carries an equal and opposite electrical charge to the proton. This was confirmation of his 1897 speech, and it was a bombshell dropped in the scientific world – the atom is not indivisible. The new particles were christened electrons, and it was they which had first been observed as cathode rays emitted inside vacuum tubes.

Although essentially a mathematician, Thomson became the head of the Cavendish Laboratory in Cambridge from 1884, a serendipitous choice as he proved uncannily good both at devising his own experiments, and deciphering what was going wrong with other people's. However, he was notoriously clumsy around the laboratory itself, so that his researchers liked him out of the way whenever possible! The Cavendish under Thomson proved a lure for bright young men in a hurry, and seven of these working at one time under Thomson were to receive Nobel prizes; Thomson also received one on his own account. Even when they did not make the breakthrough themselves, they were very quick on the uptake of other people's discoveries, notably in the cases of X-rays and radioactivity.

Note that Nobel prizes were first awarded in 1901, for physics, chemistry, physiology or medicine, literature and peace. Each constitutes a considerable cash sum, left in the will of the Swede Alfred Nobel (1833-96), the man who invented dynamite, and also owned Bofors, the arms manufacturer, and apparently felt guilty about it.

It was a Crookes tube which caused the next sensation in the new world of rays: X-rays. These were discovered accidentally by the German physicist Wilhelm Röntgen (1845-1923) in 1895, when he was 50 years old and well past the age when most of our scientists first made a mark. Roentgen was passing a current into a Crookes tube to generate cathode rays, but his tube was completely covered with black cardboard. To one side he had set up a fluorescent screen (painted with barium platinocyanide) for other purposes. He knew that the cathode rays could cause the screen to fluoresce, but was amazed to find that it did so when the glow from the cathode rays was completely covered over. Something else was getting through the cardboard – a new kind of ray. Experimenting, he allowed the rays to pass through his wife's hand onto a photographic plate. They did so, but in such a way as to show the bones of her fingers, and her ring, but not the flesh. The news of this discovery whizzed around the world, as the medical implications were obvious. Within weeks Röntgen was demonstrating X-rays to the Emperor Wilhelm II of Germany (Kaiser Bill) himself.

Nobody knew exactly what the new rays were, but they were clearly not cathode rays (streams of electrons). They were produced when the cathode rays hit the anode or wall of the Crookes tube, and then spread out in all directions. They travelled in straight lines and affected photographic plates, and they were not deflected by electric or magnetic forces – so far, they behaved like light waves. However it was not at first found possible to reflect or refract them. Only after 1910 did it become clear that they do reflect and refract, given the right type of target, and that they are indeed electromagnetic waves. Their wavelengths are very short, beyond ultraviolet light. Despite the theoretical uncertainty, as early as 1896, less than a year after their discovery, the American inventor Thomas Edison had developed an X-ray machine for medical use. What is more, the discovery of X-rays led straight to the discovery of another and yet more mysterious form of radiation.

The next step forward came from the Frenchman Henri Becquerel (1852-1908), a man who was born with something like a scientific silver spoon in his mouth. Both his grandfather and his father held the post of professor of physics at the French Museum of Natural History; Henri succeeded his father, and was in turn succeeded by his own son; so that this particular chair was held by a direct line of Becquerels all the way from 1838 until 1948.

When he heard about the new rays in 1896, Becquerel decided to investigate phosphorescent materials, which glow in the dark, to see if they too emitted X-rays. However, these materials require exposure to the Sun before they will glow, as if they need charging up. His method was to wrap up a photographic plate in thick black paper, then place it over a dish of phosphorescent salts, with some solid object such as a copper cross in between. He would then wait for an impression of the cross to form on the plates, which in some cases it did – so it looked as if phosphorescence was some form of X-ray. However one day he forgot to charge up his salts in the Sun first, but found that an impression still formed on the plate! The material in question was a uranium salt, which appeared to be producing energy from nowhere, in direct contradiction of the law of conservation of energy. At first this caused little stir in the scientific world, because the new rays looked like another form of X-rays; but Becquerel demonstrated that the new rays could be deflected by magnets, and so must be made of charged particles, and not light rays.

Becquerel moved on to other fields while the detailed study of the new rays was taken up by a girl from Poland, Marie Curie (1867-1934) and later her French husband Pierre Curie (1859-1906). Marie Curie became the first real woman scientist to achieve fame on her own account, and she was the first to win a Nobel prize (for physics in 1903 – and again for chemistry in 1911, for basically the same project). There was at the time a strong cultural connection between Poland and France. Poland had disappeared from the map a hundred years previously, being shared out between Austria, Prussia and Russia, which got the lion's share, including the part where Marie was brought up. She somehow scraped together enough money to enroll at the Sorbonne in 1890, and married Pierre, the son of a doctor, in 1895. Pregnancy intervened and it was only in 1897 the she settled down to begin a PhD on "uranium rays".

Within a year Marie had discovered that pitchblende, the ore from which uranium is obtained, is actually more radioactive than uranium itself – how could that be? It must contain some other radioactive material, but that was hard to isolate. Pierre abandoned his own research to help her to find it. By 1902, they had found not one, but two new elements, named polonium (atomic number 84) and radium (atomic number 88). Only a tiny amount of radium had been obtained – one gram – from tons of pitchblende, but it was enough for it to be analysed and found a place in the periodic table.

Pierre found that his single gram of radium was so radioactive that it could heat a gram and a third of water in an hour, then do the same thing again and again, once again in defiance of the laws of the conservation of energy – something from nothing. It was to fall to others, notably Rutherford, to determine what exactly this form of radiation was (see Chapter 12).

However, in 1906 tragedy struck – he was knocked down and killed by a carriage after he slipped whilst crossing a street in Paris. It seems likely that his fall was caused by one of the dizzy spells he had been experiencing – brought on by radiation sickness, no doubt. Marie herself lived on until 1934, when she died of leukemia, again probably due to exposure to radiation. Her notebooks are still so radioactive that they are kept in a lead-lined safe.

\*\*\*\*\*\*\*\*\*\*\*\*\*\*\*\*\*\*\*\*\*\*\*\*\*\*\*\*\*\*\*\*\*\*\*\*\*\*\*\*

The latter part of the nineteenth century saw some striking developments in a brand new area of science, microbiology. The achievement of microbiology was to establish the germ theory of disease – that diseases are caused by microscopic organisms. Up until the middle of the nineteenth century, even the most respectable of scientists such as Charles Darwin thought that these common and frequently fatal diseases were caused by bad air, known as miasma. There was plenty of this around, especially in foggy and insanitary cities like London, but it was not the cause of the diseases. These were carried by bacteria, often in bad water rather than bad air.

Microbiology is intimately connected with immunology, because having identified the agent of the disease, the next thing is to create a vaccine for it, but in fact immunology came first. The "father" of this branch of medicine is considered to be Edward Jenner (1749-1823), the man who invented the smallpox vaccine, and of whom it is said that his work saved more lives than that of any other man. Born in Gloucestershire, he worked as a country doctor, and was well aware of the danger of smallpox and of the various attempts to find a vaccine which already existed. However, he listened to his patients, and the milkmaids amongst them told him that any one of them catching cowpox was subsequently immune to smallpox, or at least, if they caught it, it was not fatal.

In 1796 he decided to test out this observation on an eight-year old boy, the son of his rather nervous gardener, a Mr Phipps, injecting him with pus scraped from the blisters of a milkmaid who had contracted cowpox from an animal called (believe it or not) Blossom. The boy had a bad day but soon recovered. Jenner, who would certainly be wondering just how much this would cost him by now if it went wrong, then tried to infect the boy with smallpox itself, but he proved to be immune, just as the milkmaids had said. Jenner published his work and began to pursue its implications full-time, aided by a large government grant of £30,000. Despite some nervousness amongst a population a little less bold or confident than the gardener Phipps, smallpox was so greatly feared that immunization spread quickly, not only in England but also in Europe. In 1803 a Spanish expedition set off for South America with a three-year program of mass immunization, the first ever such exercise.

The first name in the field of microbiology itself is that of a Frenchman, Louis Pasteur (1822-1895) who came from Dole in the Jura Mountains. Being French, he became interested in the fermentation process, and established that this is caused by micro-organisms and not, as had previously been thought, by some kind of spontaneous generation, demonstrating the maxim that all life comes from some previous life. His experiments showed that brews which are not exposed to any air do not ferment, but begin to do so when they are exposed to the air, which must then have carried with it the fermentation agent. It was clear that such agents could spoil beer, wine and milk, and in 1862 Pasteur invented his heat treatment system to kill off bacteria and moulds in these liquids, a process known ever since as pasteurization. He went on to create the first vaccines for anthrax and rabies.

The work of Pasteur came to the attention of an English surgeon, Joseph Lister (1827-1912), in particular his suggestion that gangrene might be prevented by the use of chemical solutions (without saying which). Lister, who worked as a surgeon at Glasgow University, conducted his own experiments and found that carbolic acid (or phenol, a substance derived from coal tar) did indeed have an antiseptic effect. He took to spraying his instrument and dressings with this substance and found that it was effective. In one case in 1865, an 11-year old boy had sustained a compound fracture after a cart ran over his leg. Lister applied carbolic acid to his dressings and was amazed to find that after six weeks, the bones had fused back together without suppuration. He published his findings in *The Lancet* in 1867 and this date can be regarded as the starting point of sensible antiseptic treatments.

Pasteur is regarded as one of the founding fathers of microbiology, as is the next leading name in this field, Robert Koch (1843-1910), who was born in the Harz Mountains in the Kingdom of Hanover, and was young enough to serve in the Franco-Prussian war of 1870. His first success involved the anthrax bacillus – a bacillus being a type of bacterium which produces spores which can lie dormant for long periods. He invented methods of purifying the bacillus from blood samples, so that he could grow pure cultures. He established that the spores – technically endospores – could survive in the soil, causing "spontaneous" outbreaks of anthrax amongst cattle, apparently unrelated to previous outbreaks. He published his results for this work in 1876.

He was rewarded for his work with a good government job in the Imperial Health Office in Berlin, where he continued to improve his methods of purifying his samples and growing cultures. These were grown in petri dishes, invented by and named after one of his assistants, Johannes Petri. In 1882 he found the bacterium which causes tuberculosis, the cause of one in seven deaths in Europe. All infectious diseases of this type had been thought to be caused by bad air, or miasma, but this proved not to be the case. However, Koch did not manage to find a cure for TB. The following year he identified the bacterium which causes cholera, although experimental proof of this was beyond him.

Koch is also famous for his "postulates", the rules which establish that a micro-organism is responsible for a specific disease; one of these is that it must be prepared and maintained in a pure culture. His legacy is that his pupils managed to identify the bacteria which cause a whole string of fatal or at least very unpleasant diseases, including pneumonia, gonorrhea, diphtheria, typhoid, meningitis, bubonic plague, syphilis and tetanus, all using the methods which he had pioneered – quite an impressive list! Of course, identifying the cause of a disease was one thing, but finding a way to combat that disease would be quite another.

Another contemporary if slightly later German immunologist was Paul Ehrlich (1854-1915). He conceived the idea that a compound of arsenic might be effective against one of the great scourges of the times, syphilis. He turned out to be right – there were thousands of potential compounds, but his team found the answer after testing over six hundred of them. The chemical, known as Salvarsan, represents the foundation of chemotherapy. Syphilis quickly disappeared from society. Ehrlich also coined the term "magic bullet", meaning a preparation which would knock out harmful bacteria whilst leaving the untainted tissues around them unaffected.

\*\*\*\*\*\*\*\*\*\*\*\*\*\*\*\*\*\*\*\*\*\*\*\*\*\*\*\*\*\*\*\*\*\*\*\*\*\*\*\*\*\*\*\*\*\*\*\*

This era also saw the invention of the motor car or automobile, something which was to have the greatest possible impact on the modern world. It represented the convergence of several technologies. The most important of these was the internal combustion engine, whose invention is generally credited to the Belgian Étienne Lenoir (1822-1900) in 1860, though there had been a number of previous attempts to produce one, going back over the previous 50 years. The internal combustion engine represents a very different technology to a steam engine, as the coal is replaced in the system by a gas or liquid fuel, at first hydrogen, then coal gas (though both types of engines use pistons). Combustion or firing of a fuel mixed with air (and so containing combustible oxygen) takes place within a chamber where high pressure and high temperature can be maintained. This results in a series of mini-explosions as the chemical energy of the mixture of fuel and air is converted into mechanical energy, which is used to drive a piston, in turn attached to other moving parts. By 1870, engineers working in Austria and Germany had produced the first engines to run on petrol (gasoline).

A second technology was the invention of industrial-scale vulcanization to turn rubber, which is a natural product, into more useful products such as tyres by the addition of sulfur or other "curatives". These modify the natural polymer by creating bridges between individual polymer chains. The result is a much stronger and more durable product than raw rubber. The invention is generally credited to the New Englander Charles Goodyear, who obtained an American patent for it in 1844.

Other inventions required before the first true motor car could be produced included a workable magneto or ignition system, and a carburetor for the efficient blending of fuel and air.

The German Karl Benz (1844-1929) is generally considered to be the man who produced the first automobile, though again, there had been many others working along the same lines. This first automobile was really nothing more than a horse-drawn carriage with three wheels, but fitted with an internal combustion engine. It first saw the light of day in 1885 in Mannheim, Germany  Benz's first vehicle designed from scratch to be a motor car rather than a motorized horse carriage appeared in 1889.

Less than twenty years later, the lessons learned in motor car manufacture were applied to the first aeroplanes. The Wright brothers, Orville (1871-1948) and Wilbur (1867-1912) launched their plane at Kitty Hawk beach, North Carolina in 1903. They invented the first powered flying machine because instead of trying to develop more powerful engines – in essence, the easy part – they first solved the problem of how to control an aircraft by building a series of gliders. Their great breakthrough was the invention of three-axis control, which enabled the pilot to maintain stability and to steer the aircraft. Before this the three movements of an aircraft in the air, known as pitch, roll and yaw, had proved beyond the engineers of the day. Pitch (up and down movement) was controlled by a pair of mini-wings at the front of the aircraft, called forward elevators. These must generate enough lift to enable the aircraft to take off, and in a modern aircraft consist of the horizontal flaps on the tail (i.e. rear elevators). Roll, the tendency of an aircraft to dip its wings one way or another, was controlled by wing warping, a system of ropes and pulleys to bend the wings. These enable the aircraft to bank – essentially to fly as a bird does when turning. In a modern aircraft this function is performed by the ailerons, the horizontal flaps in the wings.  Finally yaw (side to side movement or simply steering) was controlled by a rudder – again in a modern aircraft, the vertical flap in the tail.

Only when the Wright brothers thought that the controls were in place and working did they fit an engine to their plane, the Flyer I. This required them to design propeller blades, another aeronautical achievement, as propeller design is far from straightforward.  The brothers used a wind tunnel for their experiments.  Finally on 17 December 1903, flying into a freezing wind, the completed aircraft took off in a series of short flights, the longest 262 m/852 feet over 59 seconds, before crashing beyond repair. In fact many early Wright models did stall and crash. Nevertheless there was no going back – and there was an immediate interest from the military. (Wilbur Wright died of typhoid fever at the age of only 45.)

# Chapter 11 – Albert Einstein

It is now time to introduce the world's most famous physicist, Albert Einstein (1879-1955), the archetypal mad scientist, sockless but complete with electrified hair. He was born into a German Jewish family at Ulm on the Danube, where his father and uncle ran an electrical engineering business, which they eventually moved to northern Italy. Staying on in Germany for a time, young Albert somehow engineered his own removal from the high school (gymnasium) and turned up in Italy at the age of 16. Once there he renounced his German citizenship, apparently to avoid military service. He then set his sights a high-ranking technical university in Switzerland for his degree course. This was the Eidgenössische Technische Hochschule, or ETH (Federal Institute of Technology) in Zurich. A previously little-known institution in the history of science, at least it had its own entrance examination (today it boasts 21 Nobel prize winners, including Einstein himself). This allowed Einstein to apply without any qualifications from the gymnasium. He passed the examination at the second attempt, having returned to school in Switzerland for a year beforehand.

Einstein attended the ETH between 1896 and 1900, emerging with grades which were not good enough to obtain any kind of junior university post or place on a PhD program. He had been busy having fun – amongst other things, he managed to get his girlfriend pregnant (one Mileva Maric, whom he later married; the baby was given away). He is said to have got by on the minimum amount of work, and was in fact described by one of his tutors, Minkowski, as a "lazy dog...who never bothers with mathematics at all". As his professors clearly saw him as temperamentally unsuited to any form of serious application, Einstein was obliged to take work outside of the university system at the Swiss patent office in Bern, where he became a technical examiner, third class.

As it was his intention to re-enter academia, Einstein pursued his own private studies with the intention of submitting a PhD thesis to the ETH. Without a PhD, seen then as now as the minimum requirement for most entrants to university staff positions, he would have been condemned to the patent office at best. However (again, then as now) the mere possession of a PhD was still no guarantee of a university job. Einstein would have to come up with something good – so he set out to prove once and for all that atoms are a reality, and not merely a heuristic device. He did this in two ways, firstly in his PhD thesis, and secondly by a paper on Brownian motion published in the same year.

In his PhD thesis, he made an attempt to define the size of molecules. There had already been several attempts at this, including one by Thomas Young, whose estimate (for water molecules) was ten time too large. Einstein took a different approach by using the process known as osmosis. His idea was to have a container full of water, with a membrane separating this into two halves. The membrane is perforated with holes just large enough to allow a water molecule to pass through. He then considered what would happen if sugar was added to one side, making a solution in the water. However, the sugar molecules are much too big to pass through the holes in the membrane. What would happen to the level of the water on either side?

The answer is that the level rises on the sugar side. This is due to the operation of the second law of thermodynamics, whereby nature tries to iron out the differences between the two sides. By a process of osmosis (seepage), pure water seeps into the sugar solution in the attempt to dilute it to an average level. This causes the level to fall on the pure side, and rise on the sugar side. This process continues until the extra pressure on the sugar side is enough to balance the pressure of the water trying to penetrate the barrier (the osmotic pressure) Therefore the osmotic pressure can be determined by measuring the height difference. The osmotic pressure can then be increased by adding more sugar. This is the basis of the experiment, which then involves complicated calculations of the type used in the kinetic theory of gases, including the mean free path. The upshot was that Einstein was able to obtain a value for Avagadro's number – the number of molecules in a cubic centimetre (or mole), now known to be $6.022 \times 10^{23}$. In 1905 Einstein's value was $2.1 \times 10^{23}$, revised within six years to $6.6 \times 10^{23}$ – his number was three times too small at first, but of the right order of magnitude. A problems which had been outstanding for a century had, in effect, been solved by a part-time PhD student. The examiners accepted his thesis in 1905.

In this *annus mirabilis*, 1905, Einstein also published four papers, one of them on the subject of Brownian motion, in his second attempt to pin down the existence of molecules and so atoms. This is named after a Scottish botanist called Robert Brown (1773-1858). When studying pollen grains under a microscope in 1827, Brown observed that these tiny grains, which typically have a diameter of less than .005 mm, move around in a jerky fashion. He also observed the same phenomenon on purely inanimate particles. Others had looked at the problem, but Einstein started from first principles. He decided that it could be approached statistically.

It seems that the particle is being bombarded from all sides by collisions with individual molecules, but as it is jerked in one direction, the next jerk, in terms of probability, could be back to where it started, or in some other direction. The end result is a zig-zag path in which the total distance a particle travels from its starting point – every zig and zag added together – is proportional to the square root of the time from the first movement. He calculated that a particle should move 6 thousandths of a millimetre in one minute, twice as far in four minutes, and four times as far in sixteen minutes. The elapsed time grows greater to move any distance, because of the square root function.

Einstein's paper included a precise mathematical description of the problem which related it to the total quantity of molecules present (as given by Avagadro's number). This work did not come out of thin air – J J Thomson had also contributed to the field, in lectures given as recently as 1903, and an equation found in the work of the German Nobel prize winner Walther Nernst from 1888 found its way into Einstein's paper. Nevertheless the paper was very influential and its methods have since been used in other ways, such as in the measurement of radioactive decay. Verification of Einstein's equations involved accurately measuring the slow drift of particles, and could not have been easy, but it was achieved by the Frenchman Jean Perrin (1870-1942). Partly because of the importance attached to the newly-established reality of atoms and molecules, Perrin got the Nobel prize for his work, and Einstein was very grateful to him.

Einstein's next effort from 1905 gave us his Special Theory of Relativity, in a paper entitled *On the Electrodynamics of Moving Bodies*. This appeared without any citations or indications as to what had preceded it; it contained no experimental results and almost no mathematics – in other words it was purely a thought experiment – yet one which changed the world.

This piece of work emanated from Maxwell's equations and the subsequent attempts to test for the ether wind in the Michelson-Morley experiment. A Dutch physicist called Heindrik Lorentz took the view that this experiment had failed, and he produced what are called the

Lorentz transformation equations in 1904 to indicate why that was so – because the Earth shrank by a tiny amount in the direction of motion. This was not correct, but it must have got Einstein thinking, because his equations for Special Relativity were mathematically identical, but proposed a different solution.

Einstein started from the point made by Maxwell's equations, that the speed of light $c$ is always constant. And what had the Michelson-Morley experiment demonstrated? That the speed of light is constant, no matter how it is measured between different points on the surface of the Earth. However, this flies in the face of Newtonian mechanics. According to Newton, the speed of light should depend on the relative speed of the observer of the light. For example, if I am in a car travelling at 30 miles per hour, and another car approaches me at 50 miles an hour, then the other car is travelling at 80 miles an hour relative to me. If a third car later passes me going at 40 mph, and I am still going at 30 mph. then its speed relative to me is 10 mph. But Maxwell's equations said that as far as light is concerned, this is not the case. Its speed is always the same no matter how fast a body is approaching or receding. If say a comet pitches into the Sun going at half the speed of light, then the light travelling from the Sun is still observed as the standard 300,000 km per second from the point of view of a terrified observer on the comet.

It is a logical consequence of this that if speed is constant, then time and distance (space) must be variable. Time on the comet will only be passing at half the speed of someone observing it as it crashes towards the Sun. If only it could reach the speed of light, time for the observer on the comet would stand still altogether! However, Einstein predicted something else for that comet. At that kind of speed, it would have grown considerably larger in mass! In fact at the speed of light the comet would become infinitely large, which means that at that point any equations become irresolvable (meaning they have no solution – usually the point at which mathematicians have to give up and think of a new theory). In practice, this means that nothing can travel as fast as, or faster than, light.

Einstein's remarkable new theory also led to his most famous equation, $E = mc^2$. (This did not appear in the paper itself, but in a short supplement issued several months later.) This states the equivalence between energy and mass. According to this, mass can be CONVERTED into energy by multiplying it by the speed of light, squared. As the speed of light is 300,000 km per second, it is immediately plain that a very small amount of mass converts into a very large amount of energy. This is the principle behind the atom bomb. It immediately showed why the Curies had been able to generate an apparently inexhaustible amount of energy from a tiny piece of radium, and it soon became obvious to the geologists and astronomers that it also expanded the supplies of energy available to the Earth and the stars into timescales of billions of years. From the emergence of this equation the discoveries of Einstein begin to seem almost supernatural, in the manner of Newton. That there may be a relationship between energy and mass seems fair enough, but what could the square of the speed of light possibly have to do with it?

The theory is called "relative" because Einstein saw that all motion is relative, and depends on the position of the observer, say A (on Earth) observing B (on the comet), or B observing A. It is called "special" because it deals with a special case, where all observers are moving at constant velocities relative to one another (so each one thinks he is at rest, and everything else is moving around him). The more normal case where one or more observer is accelerating was more difficult to express and in fact it took Einstein until 1915 to present this to the world as the General Theory of Relativity.

The Special Theory did not refute Newton's mechanics, but built on them. What it really said was that the common sense of the Newtonian system was all very well for the sedate speeds observable in the orbits of planets and so on, but it broke down at speeds approaching the speed of light. Time and space must vary to accommodate a speed of light which is constant regardless of the motion of the observer. Bizarre as it seems at first, the Special Theory has been tested and proved experimentally, for example using beams of particles accelerated to speeds approaching the speed of light, and it works.

One person who was watching Einstein's progress very carefully was his old tutor at the ETH, Minkowski – in fact the man who had described him as a "lazy dog". He made Einstein's work much more understandable in 1908 when he fitted it into a four-dimensional geometry which made the difference between the Lorentz transformations and the Special Theory clear. The four dimensions were the three standard ones (Cartesian $x$, $y$ and $z$) plus time $t$. In other words space and time were now fused into space-time. Minkowski found that he could specify the location of an object in space-time by adding $x$, $y$ and $z$, and subtracting $t$!

Einstein's fourth paper of 1905 was on the subject of light quanta, and although only short, it introduced such a revolutionary idea that is was for this work that Einstein was to receive his Nobel Prize in 1922.

To understand this subject we have to go back five years to another German physicist called Max Planck (1858-1947). Planck's research area was black body radiation. A black body is one which absorbs all radiation, but the puzzle concerned the nature of the radiation FROM the black body when it is heated. In this case it appears independent of whatever the black body is made of, and dependent only on the temperature. In practice, a lump of iron acts very much like a black body, and glows red when hot, and yellow when hotter still. This relationship can also be measured in the stars, where it is used to assess their temperature.

Nobody could find a formula which would express black body radiation mathematically until Planck eventually did so, but only by dividing the radiation into tiny but discrete particles – and of course, everyone knew that radiation (of red of yellow light) was a wave, and not a series of particles. The formula to calculate the number of these particles included a constant, $h$, known after Planck's announcement in 1900 as Planck's constant. However this constant was only there as a kind of placeholder until the mystery of black box radiation was solved. Nobody thought that discrete packets of energy or light, known later as quanta, were real.

In the meantime there had been a number of studies of what is called the photoelectric effect, in which ultraviolet light is observed to knock out electrons from the surface of elements with a high atomic number. No matter what the strength of the light, the electrons were found always to have the same energy. There would be more of them with stronger light, but they would still each have the same energy.

The start of the quantum revolution in physics is sometimes dated to Planck's announcement in 1900, but the real breakthrough came with Einstein five years later. He calculated that radiation from a black body would always be emitted in discrete units which were in fact multiples of $hv$, where $h$ is Planck's constant and $v$ is the frequency of the radiation. The photoelectric effect could also be explained if the light shone on the receiving metal consisted of a stream of individual particles, or light quanta. These would then impart the same amount of energy to an electron in the metal, which is the reason the emitted electrons all have the same energy. The photoelectric effect could not be explained by the wave model of light. Einstein concluded that light consists of a finite number of energy quanta, absorbed or generated as discrete units. These light quanta were later given the named photons.

This was a revolutionary idea at the time – hadn't Young's double slit experiment demonstrated conclusively that light is a series of waves? This was the start of the realization that light sometimes behaves like a wave, and sometimes like a stream of particles (as originally envisaged by Newton) – what is called wave-particle duality.

Einstein's suggestion so infuriated an American professor called Robert Millikan (1868-1953) that he spent the next ten years devising experiments to disprove it, only to conclude that Einstein had been right all along.

Einstein's papers all appeared in a German physics journal called *Annalen der Physik*, which allowed publication despite the fact that the author had no university affiliation. It was not until 1909 that Einstein was finally able to quit the patent office and return to the universities. By 1914 he was a professor in Berlin. Meanwhile, he had married Maric in 1903, and produced two more children. By 1914 the couple were living apart, and were divorced in 1919 – such is fame! However by 1915, Einstein was ready with his General Theory of Relativity. This was to be a giant leap forward from the special theory. It is thought likely that if Einstein had not published his special theory in 1905, then someone else would have had the same idea within a few years – but how long would it have taken the world without Einstein to produce the general theory? Perhaps a generation.

One reason that the development of the theory took so long was that Einstein needed to master more mathematics first. In particular this involved the mathematics of curved surfaces, known as non-Euclidian geometry, as expounded by nineteenth-century mathematicians including Bernhard Riemann. If only he hadn't been such a lazy dog at the ETH!

The general theory deals with bodies which are accelerating, not just moving in straight lines at constant speed. Unlike the special theory, it takes gravity into account. It considers the question – what effect does gravity have on moving things, and especially on light? Einstein's insight was to see that there is no real distinction between acceleration and gravity. Apparently this idea came to him whilst sitting at his desk in the patent office, when it occurred to him that a man falling off the roof would feel quite weightless on his way down! The acceleration of the downward movement would cancel out the feeling of weight exactly (so would the feeling of weight gradually diminish as the acceleration increased?)

Einstein took the view that the universe is flat, but this flatness is distorted locally by the presence of the stars, which bend it. Gravity is seen as the interaction between space-time and matter, or the distortion of flat space-time by mass. Matter bends space-time in its locality, but space-time dictates the movement of the masses within it. The orbit of a planet round the sun is put in a different way – the planet is actually following a straight line in curved space-time. The curvature is due to the gravitational influence of the Sun. Another way of putting this is to say that the gravitational field of force round a large object like the Sun is replaced by a curvature of the otherwise flat space-time. This curvature replaces the idea of a mysterious force (gravity) operating in a vacuum between two objects – something which Newton himself had distrusted.

Einstein's equations had one bizarre and (to him) totally unexpected feature – they needed space to be either expanding or contracting; otherwise they did not work. This was another example, like the Maxwell equations, of mathematical logic showing up a yet-to-be realised physical reality. So Einstein added another term, known as the cosmological constant, to allow for the static universe he thought lay around him. He was later to call this his greatest blunder, because within a few years Hubble had proved that the universe is indeed expanding. (There have been later attempts to revive the cosmological constant.)

So does the theory mean that time travel is possible? In theory, yes. If a spaceship left the earth and travelled for 100 years at half the speed of light, the time elapsed on board would be much less than the time elapsed on earth, as its clock runs more slowly. So when it came back, it would arrive much more than 100 years in the future – exactly how many depending on its average speed. However, to travel backwards in time in not thought a practicality by modern scientists.

Though widely publicised as headline news, Einstein's general theory was held to be too complicated for the layman to understand. In fact when the British astronomer Sir Arthur Eddington was asked if it was really the case that only three people in the world understood it, he replied "I am trying to think who the third person is!"

The reality of relativity was dramatically confirmed in the solar eclipse of 1919. The theory predicted that light from distant stars would be slightly bent by the gravitational field of the Sun. This effect could only be observed during an eclipse when the Sun's brightness does not block out the light from the stars behind it. All that would be necessary would be to take photographs of the stars in the normal night sky and then again during an eclipse, and compare the two. The Royal Astronomical Society in London organized expeditions to the West African island of Principe and to Brazil, under Arthur Eddington, to photograph the eclipse and test for the effect. When the results were reported confirming Einstein's theory, they caused headline news around the world. They did not show, however, as some claimed, that light has weight. Light consists of electromagnetic waves which are affected by gravity, even though gravity itself is not an electromagnetic force. In fact light waves are a form of energy, and as far as gravity is concerned, mass and energy are the same thing.

The results of Eddington's expedition transformed Einstein from obscurity to the status of an international celebrity. Never had a single scientist achieved such fame. In fact, Eddington was already the champion of Einstein's theories in the English-speaking world, and there have been subsequent attempts to discredit his findings, but no one seriously questions the theory nowadays.

Einstein remarried in 1919, this time to his cousin Elsa Löwenthal, who was to die young in 1936. In 1933, whilst Einstein was visiting the USA, Adolf Hitler and the Nazi party came to power in Germany on a vicious anti-Jewish program. Einstein and his wife got back as far as Belgium, where they learned that his cottage had been raided, that Jews had been forbidden from holding official positions including any university post in Germany, that his books had been burned by the Nazis, and that there was a bounty of $5,000 on his head. Unsurprisingly, he never returned to Germany. By October he had been offered a wonderful job at an Ivy League university in the USA, Princeton. When he arrived there he said he thought he had gone to heaven.

# Chapter 12 – Early Twentieth Century Chemistry and Physics

Following the Curies, the next person on the trail of radiation is Ernest Rutherford (1871-1937), considered the greatest experimenter since Faraday. His parents had emigrated from Scotland to New Zealand to raise a little flax and a large family. By1895 he had found his way to the Cavendish Laboratory in Cambridge. Because of a quirk in the system of rules for research students at Cambridge at that time, he was obliged to move on in 1898, when he went to McGill University in Montreal.

In his experiments on radiation, Rutherford identified three different types of rays, which he named alpha, beta and gamma. It eventually emerged that alpha rays are the same as helium atoms without the electrons (that is, cations of helium, the second-lightest element, which normally has two protons and two electrons). Alpha rays only have a short range, and can be stopped by a sheet of paper or even a few centimetres of air. So what happens to their electrons? Enter the beta rays, which are streams of energetic electrons with a much longer range and penetrative power. The gamma rays, discovered later by Rutherford, are light (electromagnetic) rays with a very short wavelength, even shorter than X-rays. All three types of ray are classed as ionizing, that is, strong enough to separate an electron from an atom, and if that atom happens to be sitting in human flesh, the results can be fatal. The most damaging of the three are the gamma rays, followed by the beta rays.

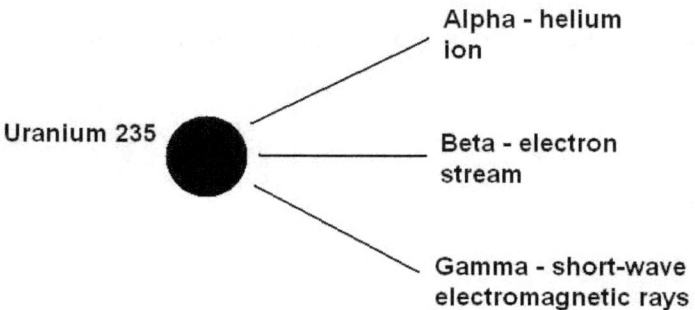

Whilst in Montreal, Rutherford struck up a partnership with an English scientist, Frederick Soddy (1877-1956). This pair discovered that during the process of radioactive decay, an atom of one element is converted into the atom of another – had they found the philosopher's stone at last? No, the transmutation was not from lead to gold, though lead is involved! Uranium will eventually decay into lead, giving off helium as it does so.

The pair established that radioactive elements decay at set, measurable rates. For example, radium loses half its atoms over a period of 1602 years. This, then was the source of its seemingly inexhaustible energy supply as noted by the Curies. After another 1602 years, another half of the radium has decayed. This is the "half-life" of a radioactive element. As the half-life of radium is evidently rather short, it cannot have been around since the formation of the Earth – in fact it is created by the radioactive decay of uranium, which has a much longer half-life (4.47 billion years in the case of uranium-238, and 704 million years in the case of uranium-235). Rutherford was quick to point out that here was a source of energy which could make the Earth itself very old indeed. He calculated the age of a lump of pitchblende, the principal ore of uranium, as 700 million years. In 1904 he took the rock with him to a lecture which he gave at the Royal Institution in London. He was unable to shift Lord Kelvin, the great authority whose latest estimate of the age of the Earth had come down to a mere 24 million years, but the rest of the world took due notice. Here at last was evidence for what Darwin had been saying all along, that the timescale for evolution was to be measured in hundreds of millions of years. Rutherford had given the geologists and the biologists time, lots and lots of it.

Rutherford returned to England in 1907, this time as Professor of Physics in Manchester (funny how often that place crops up!), where the research facilities were outstanding (and moreover, J J Thomson was still in charge at the Cavendish). Here he was able to use alpha particles to probe the structure of the atom itself, in association with his assistants, Hans Geiger (1882-1945) and Eric Marsden (1889-1970). It was of course Geiger who developed the eponymous Geiger counter, a detector which picks up the radiation of alpha rays.

Until this time, it was thought that an atom consisted of a sphere of positively charged material in which the negative electrons were somehow embedded, like the pips in a melon, but the new teams' experiments showed otherwise. They fired alpha particles at a sheet of gold foil, and found that most of them passed straight through, though some were deflected, and others bounced straight back. Evidently the atom was not as solid as previously thought, and must in fact consist mainly of nothing, as a helium cation could pass straight through! Those few which bounced back had evidently hit the (positive) nucleus of a gold atom head-on and been repelled backwards. The experiments eventually showed that the nucleus occupies only one hundred-thousandth of the diameter of an atom, sitting like a grain of sand in the Albert Hall. It is surrounded by a web of electromagnetic forces which link it to the electrons.

Rutherford went on to find that nitrogen atoms when bombarded with alpha particles convert into oxygen, with the emission of a hydrogen nucleus. Hence he achieved the first artificial transmutation of one element into another. By now, using his alpha rays, he had completely transformed the conception of the atom. It has a central core, the nucleus, which contains positively-charged particles, which he called protons. This was surrounded at a great distance (on the atomic scale) by a cloud of almost infinitesimally small but negatively charged electrons.

Rutherford eventually found his way back to the Cavendish Laboratory, where he succeeded J J Thomson (another Manchester man) as head in 1919. Not a man to mince his words, Rutherford thought that only physics was "real science" – the rest of it mere "stamp-collecting". Many contemporary biologists may secretly have agreed with him, but few were willing to say as much.

Of course this means that our old friend J J Thomson was still active in 1912, and in that year, along with his assistant Francis Aston (1877-1945) he invented the mass spectrometer using a development of the instrument he had previously used to measure the ratio of electrical charge to mass ($e/m$). The idea of the mass spectrometer is specifically to measure the mass-to-charge ratio ($e/m$ or $m/e$) of a molecule.

Thomson and Aston sent a beam of ionized neon gas down a tube through a magnetic and electric field. As noted, an element which has been ionized has had one or more of its electrons removed. They observed that the ionized neon created two (instead of one) patches of light on a photographic plate. They concluded that this meant that there must be two types of neon, now called neon-20 and neon-22. Frederick Soddy had already suggested that atoms might come in different flavours, which he called isotopes, and here was experimental evidence. Aston got his Nobel prize for this work in 1922 (Soddy had got one in 1921). However they still did not know WHY there should be two or more varieties of an atom.

This did not become apparent until 1932, when the neutron or neutral particle within the nucleus was discovered by James Chadwick (1891-1974). Chadwick picked up on work being done in Paris by Frederic and Irene Joliot-Curie. This pair had bombarded beryllium with alpha particles, finding that the resulting radiation ejected hydrogen atoms from paraffin (which contains rather a lot of them). They thought that the artificial radiation coming from the beryllium was some form of gamma ray. In a reverse of the Priestley/Lavoisier story, when only the Frenchman understood what Priestley had found, Chadwick reinterpreted the results from the Joliot-Curie experiment. He conducted some more tests of his own using boron as the target, and in a few short weeks had confirmed the existence of a previously unknown particle, the neutron. (He too received a Nobel prize, in 1935). It was the neutrons which were knocking the hydrogen atoms out of the paraffin.

So what exactly is an isotope? An atom has three constituents: the positively-charged protons, at the centre; the negatively-charged electrons, "orbiting" around it; and a variable number of uncharged neutrons which form the nucleus of the atom with the protons. The number of protons and electrons in the atoms of any one element is always the same, and give the element its atomic number – for example, oxygen 8, carbon 6. The chemical properties of an element derive from this number and are not changed by the number of neutrons. The different varieties of an atom represent the variable number of neutrons. For example, oxygen has three stable isotopes because it can have 8, 10 or (very rarely) 9 neutrons (plus always 8 protons, giving oxygen-16, -18 and -17). Although they do not affect the atomic number of the atom, the isotopes affect the atomic weight. This may not be a round number at all, as it is an average of the different weights of the various isotopes.

An unstable or radioactive isotope can give off rays which transform it into another isotope of the same element, or a different element altogether. If the number of protons is affected, the result is a different element. If it is the number of neutrons, the result is a different isotope of the same element. Thus as noted, uranium 238 decays into lead and helium. This makes it suitable for geological dating. Other elements, some not normally thought of as radioactive, can also be used in radioactive dating, notably thorium, potassium, lead and carbon. Of these, carbon-14 has a half-life of only 5730 years, so carbon dating is much more useful to the archeologist than the geologist. It was invented by Willard Libby at the University of Chicago in the 1940s and enabled real dates to be ascribed to archeological and prehistoric artifacts for the first time.

*******************************************

One of the most famous scientists to work with and follow Rutherford was the Dane, Niels Bohr (1885-1962) (following in the footsteps of other Danes – Brahe, Romer, and Steno). Born in Copenhagen, he came from an academic family – his father and brother were both professors. By 1911 Bohr had arrived at the Cavendish in Cambridge, but did not get on well there, ending his stay in England with Rutherford in Manchester. He returned to Manchester during the war years 1914-6, braving the passage across the battleship-infested North Sea (Denmark was neutral) to do so. However, by that time his greatest work was already done.

Bohr had the happy knack throughout his career of knitting together whatever materials lay at hand to construct his theories. He didn't worry too much about mathematical rigour or consistency, for example, as long as his model matched experimental results. In his 1913 model of the internal structure of the atom, Bohr considered the problem of electrical charges. Why didn't the electrons simply fall into the nucleus by electrical attraction? In fact, despite the attractiveness of the analogy, electrons cannot be said to orbit a nucleus in a way that a planet orbits the Sun. The planets are kept in their places because they are moving, and this gives them enough centrifugal force to balance the gravity of the Sun. An electron, however would have to accelerate around its orbit as it changes direction within it, emitting radiation in the form of electromagnetic waves as it does so. In this case it would lose energy and fall into the nucleus in a fraction of a nanosecond; so clearly there must be some other mechanism, outside of the physics of Newton and Maxwell. Step up the next application of quantum physics.

Bohr proposed that electrons do not cloud haphazardly around the nucleus, but instead hold their places on a series of concentric rings. The innermost ring could hold a maximum of two electrons, the next eight and the third up to eighteen. Within these rings, the electrons are not completely fixed in their orbits – they may move between rings when, for example, a gas or a metal is heated or cooled. (Also, an electric current is in fact a flow of electrons. As electrons move along a copper wire, the copper may lose electrons and so become ionized, but it remains copper for all that.) Bohr realised that the number of spare slots in the outer ring of electrons of any element determines its chemical properties. If there are spare slots then the element seeks to combine with other elements in compounds.

This concept proved to be the key to the unlocking of the secrets of the way that compounds are formed. More details of the way in which this happens were worked out by others over the next few years as the principles of chemical bonding came to be understood. For example, any element which has NO spare slots in its outer ring – such as neon, which has ten electrons, two and then eight – is inert, and does not react with anything. This is why we have the noble gases – neon, helium, argon, krypton, xenon and radon. In any event, Bohr's principle that electrons have defined orbits or rings has stood the test of time.

Bohr then said that the electrons stay in their orbits because they do not have the capacity to emit radiation continuously. They can only do so in discrete packets, or quanta of light (photons). If an electron did this, it would jump down to a lower ring. However it cannot do this if the inner rings are already "full". The electrons on the innermost ring are, under the Bohr model, and for no known reason, simply forbidden to jump into the nucleus (the explanation for this came from Heisenberg ten years later).

Under this model each jump of an electron, up or down, corresponds to the absorption (up) or release (down) or of a precise quantum of energy. If a large number of atoms are radiating in this way, then a bright line will appear in the spectrum at that given wavelength. In the case of hydrogen gas, Bohr found that the spectral lines as predicted by his model matched the known signature of hydrogen gas. He had explained why each element has its own special spectral fingerprint of Frauenhofer lines.

This model, a quantum add-on to the Rutherford model, raised as many questions as it answered, but it seemed to show the way forward, and indeed it is still in use today. Because of it, Bohr became a scientific celebrity, and his own government enticed him back from Manchester to Denmark. Here he founded the Niels Bohr Institute, which, in the manner of Tycho Brahe and his observatory, drew the famous physicists of the day to Denmark. When the Nazi Germans invaded Denmark in the Second World War, Bohr (and his son Aage, also a scientist) fled to England and then on to the United States, where he advised on the Manhattan Project to build the first atomic bomb.

Meanwhile another form of spectrometer had become available to scientists to enable them to probe increasingly complex substances. This was the X-ray spectrometer, developed by William Bragg (1862-1942) and his son Lawrence Bragg (1890-1971). Between them this pair worked out the rules for the diffraction of X-rays within crystal lattices, a subject known as X-ray crystallography. They published a book on the subject called *X-Rays and Crystal Structure* in 1915, and were duly awarded a Nobel prize between them, Lawrence, at the age of 25, being the youngest person ever to win one. The new machine was to play its part in the next stages of the evolving story of the nature of electrons, but really came into its own in studies into the structure of the materials of the body – hair, muscle, skin and so on – from the 1930s onwards.

The next breakthrough in quantum physics appeared in a PhD thesis written in 1924 by a student at the Sorbonne, Louis de Broglie (1892-1987). Louis was an aristocrat – his father was a duke, and he eventually succeeded to the title as the seventh Duke of Broglie. He put forward the view that, just as electromagnetic waves can be described as particles – photons, following Einstein's paper of 1905 – all material particles can be described in terms of waves. Using an equation which had come out of the general theory of relativity, De Broglie showed that the momentum of a particle (of light) multiplied by its wavelength is equal to Planck's constant (which had been derived showing that light waves are made up of discrete quanta). De Broglie suggested that the same equation applied to electrons, which unlike light, have mass. His model for this showed electrons as waves, aptly described by John Gribbin as "running round their orbits like a snake biting its own tail". The place of the electron within the electron "shell" of the atom would then correspond to the different harmonics of these waves. A harmonic in this sense is a wave frequency a whole number of vibrations from the base frequency, as an open guitar string touched at half its length will sound the same note one whole octave higher. Only harmonic orbits (where the peaks and troughs of one wave coincide with those of the next, instead of cancelling each other out) would be allowed. There is a resemblance to the Bohr theory here, in that there can be no graduation from one state to another, but only whole-number states. De Broglie's thesis supervisor was so baffled by all this, suspecting it might just be an illusion, that he sent it to Einstein, but Einstein liked it, and indeed it has been described as the best PhD thesis of all time.

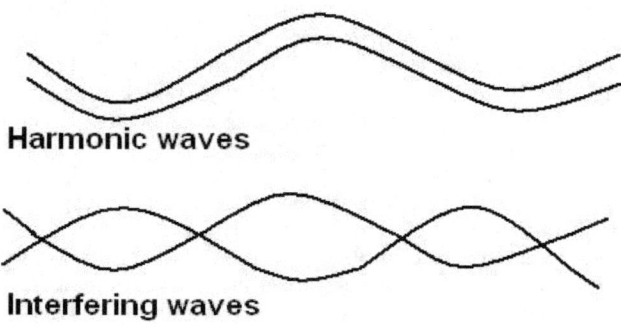

Only harmonic waves fit the electron orbit of De Broglie

To test the theory, De Broglie noted that the electrons involved should have the right wavelengths to be diffracted from crystal lattices, which have the right kind of spacing to pick up electron waves. This could be investigated using X-ray crystallography, a new tool available to scientists from 1915. Several scientists got involved with testing this idea, apparently no easy task, and they duly confirmed the theory. One of them was George Thomson, son of J J Thomson (the man who first discovered the electron), who was at the University of Aberdeen. He won a share of a Nobel prize for this work in 1937, in a nice illustration of the topsy-turvy world of quantum physics – because his father got a Nobel prize for proving that electrons are particles, and he got another for showing that they are waves; and in their own ways, both of them were right.

The waviness of matter at the molecular level is not something which would occur to anyone at everyday scales. It only has any importance in the crazy quantum world of the atom, where common sense has no place, and every certainty turns into an uncertainty. Within a few years of De Broglie, the physicists had come up with equations which equally well described electrons as waves and particles. Three of the scientists/mathematicians involved are well-known for other reasons: Erwin Schrödinger (1887-1961), Werner Heisenberg (1901-1976) and Paul Dirac (1902-1984).

Heisenberg's second contribution to the wave-particle duality debate is his famous uncertainty principle. This states that certain pairs of quantum properties, normally position and momentum, cannot both be known at the same time. Simply measuring one causes the other value to become unknown. This is a fundamental feature of quantum mechanics and not something simply brought about by inaccuracies in measuring, or an inability to make measurements. This is because the electron itself does not "know" exactly where it is, and where it is going, at the same time. The theory enabled Heisenberg to tell the world why the negative electron on the innermost Bohr-style ring does not fall straight into the nucleus. If it did, it would violate the uncertainty principle! We would know both its position (in the nucleus) and its momentum (not going anywhere)! Another more understandable way of putting this is that the wavy orbit of the electron is much too big to fit into the nucleus.

Schrödinger, an Austrian, produced an equation for the electron as a wave – and mathematicians were only too happy to deal with something as familiar as a wave equation – which, though satisfying mathematically, makes no assumption that De Broglie's waves are physical things.

At one point Schrödinger expressed exasperation at the other physicists involved in quantum mechanics. This followed the announcement of the "Copenhagen interpretation", the work of Niels Bohr, Heisenberg and others, which said that nothing is certain at the subatomic level. Hence Schrödinger invented his famous cat, shut up in a black box with a vial of poison which may or may not have killed it. According to the Copenhagen interpretation, one could not say if the cat is dead, so it must be presumed to be both 100% alive and 100% dead at the same time, which is the reason Schrödinger produced his thought experiment in the first place. His view was that from the objective, scientific point of view, the cat is either dead or alive, and definitely not both at the same time! Even Einstein struggled with the uncertainty principle, famously saying that "God does not play dice", but both

Schrödinger and Einstein were trying to apply everyday common sense to subatomic physics, where common sense flies out of the window.

Paul Dirac was an English mathematician, born in Bristol to a Swiss father and a mother whose own father had been a ship's captain. His father, a somewhat rackety language teacher (who was eventually found to have kept another woman) insisted on his children addressing him in French, which obliged poor Paul to suffer much in silence! Family life was blighted by the suicide of his brother Felix in 1925. Dirac saw the grief in his parents, and said that he never realized just how much parents cared for their children.

At Bristol university, where he studied electrical engineering, Dirac was marked out for stardom by his tutors. He found his way to Cambridge, where he eventually became Lucasian Professor of Mathematics – Newton's old job. In his twenties – always the most, if not the only, productive period for brilliant mathematicians – he managed to produce a paper which fitted Schrödinger's wave equations for the electron into the framework of the special theory of relativity (1928) – the last word in electron equations. Curiously, however, the equation had two solutions, one negative for the electron, and the other positive! Dirac's 1928 paper also contributed to the understanding of that phenomenon which is so baffling to the general reader, spin, the angular or rotational momentum of subatomic particles such as electrons.

By 1931, having been unable to resolve the two solutions into one, Dirac came to the conclusion that his own equation was in fact predicting the existence of a positive electron, or positron, a particle unknown to the world – antimatter! The evidence suggested that if enough energy is available (energy converts to matter on the lines of $E = mc^2$) then it would convert to two electrons, negative and positive. In experiments carried out in 1932-3, the American Carl Anderson (1905-1991) succeeded in creating a positron. The world of physics would never look quite the same again.

A biography of Dirac called *The Strangest Man* by Graham Farmelo was published in 2010.

\*\*\*\*\*\*\*\*\*\*\*\*\*\*\*\*\*\*\*\*\*\*\*\*\*\*\*\*\*\*\*\*\*\*\*\*\*\*

In modern chemistry, the name is bond – chemical bond. There are several different types, of which more in due course, but the first to be discovered is now named the Van der Waals force after the Dutchman Johannes van der Waals (1837-1923) who first described some of its properties in 1873. Certain materials bond strongly in two dimensions, but not in a third, so forming layers. All that holds them together in the third dimension is a mild polarity, where the top of the layer is slightly positive, and the bottom negative, or vice-versa. Despite the weakness of this force – perhaps one-fortieth of the strength of a covalent bond – it may be the only thing which gives some materials a coherent shape. Clay, for example, consists of fine sheets or lattices of the oxides of silicon and aluminium, but between these sheets are only van der Waals forces. They are very easy to break, which is the reason clay smears. Again, graphite is a form of carbon built into flat sheets, strongly bonded, but with only the van der Waals force between the sheets. This is the reason that a "lead" (in fact graphite) pencil writes so readily – this force is easy to break open.

The next type of bond to be discovered was the ionic bond, most of the credit for which is generally ascribed to the Swedish chemist Svante Arrhenius (1859-1927), who received a Nobel prize for his work on ions in 1903. The best-known example of an ionic bond is found in common salt, NaCl. Electrons "orbit" the nucleus of an atom in concentric shells, the first one with room for two, the next for eight, and the third also for eight/eighteen. Sodium has a total of eleven electrons, which means that there is only one in its outer shell, leaving seven gaps which nature tries to fill. Chlorine has a total of seventeen electrons in its cloud, first two, then eight, then seven, which leaves one vacant slot in its outermost ring. The two elements get together electrically. Sodium loses its spare electron, giving it a positive charge within the molecule, and chlorine gains that electron, giving it an overall negative charge. These electrically charged atoms, now called ions, are held together by electrical forces in a crystalline array. The reaction between sodium and chlorine to form salt is known as an exothermic one, giving off heat, because the total energy of NaCl is slightly less than that of the separate Na and Cl atoms.

There is technically no such thing as an ionically bonded molecule, because independent molecules of a substance like common salt cannot exist in the way that say molecules of oxygen $O_2$ can; instead they form crystals. Nevertheless their electrical bonds are very powerful – in fact much the most powerful of any type of chemical bond.

Another way of joining atoms together is the covalent bond, first described by the American scientist Gilbert Lewis (1875-1946) in 1916. The chemistry of life is built around covalent bonds with carbon at their centre. Carbon has two electrons in its inner ring, and four in its second ring, giving six altogether, and leaving four vacant slots. Each of these can pair up with an electron offered by hydrogen, because hydrogen has one electron and so also one vacancy on its single ring of electrons. If four atoms of hydrogen combine with one atom of carbon, by sharing their spare electrons, each has the illusion of having filled its outer ring – they have shared their valency, hence the term covalent bond. The result of this happy mixture is one molecule of methane, $CH_4$. Hydrogen gas ($H_2$), oxygen gas ($O_2$) and water ($H_2O$) are all formed in the same way. It is estimated that there are at least 100,000 different (carbon-based) molecules in the human body! The dry weight of a human being is about two-thirds carbon.

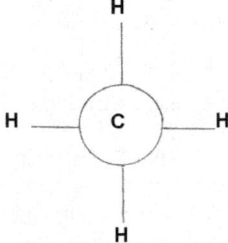

The structure of methane

Diamonds are not like graphite because the carbon atoms from which they are made are bonded covalently with other carbon using all four valencies, giving a very tight three-dimensional structure.

Except in the form of carbonates (which as calcium carbonate form limestone and chalk), carbon does not play so central a part in inorganic chemistry as it does in organic chemistry, where its ability to combine with other elements leads to the creation of huge organic hydrocarbons and other molecules. This is because it does not combine with metals. Metals themselves prefer to form ionic bonds (cation to anion, normally metal to nonmetal, for example sodium chloride as described above).

Another covalently-formed gas is hydrogen chloride, where just one hydrogen electron is needed to make up the complement of seventeen chorine electrons to a full house of 18.

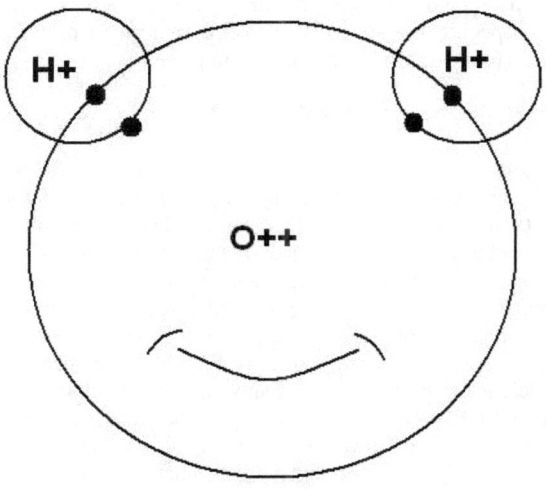

The hydrogen bond as imagined by Natalie Angier, with Mickey Mouse ears. The negative electrons lie closer to the oxygen nucleus giving the hydrogen atoms a slight positive charge.

The fourth type of chemical bond is the hydrogen bond, which was first discovered and described by several scientists working in the period 1912-20, though its properties were later described quantitatively by Linus Pauling, of whom more in due course. A hydrogen bond has about four times the strength of the van der Waals force, but still only one-tenth of the strength of a covalent bond. Nevertheless, it has the advantage of great elasticity. Water ($H_2O$) bonds covalently, but a collection of water molecules also bonds using the hydrogen bond, and it this force which makes it "sticky", so forming droplets easily. The electrical charge of the water molecule is not quite evenly distributed, so the two hydrogen atoms have a slight positive charge, and the oxygen atom a slight negative charge, as the shared electrons (which are negative) lie closer to the heavier oxygen proton. Hydrogen bonding is not confined to water, affecting other covalently bonded molecules where the lightweight hydrogen is a partner.

In the human body, hydrogen bonds perform the vital function of holding the two halves of the DNA helix together. They are strong enough to maintain the shape of the helix, but weak enough to allow the two strands to be unzipped as and when required. These bonds are also used to create the specific shapes which proteins require to perform their vital functions (for example, the haemoglobin protein).

*********************************************

James Chadwick, the man who discovered the neutron in 1932, was also involved in the identification of the strong nuclear force during the 1920s. It came to be realised that the cluster of separate positive particles (protons) in the centre of the atom, the nucleus, should simply blow apart because of the magnetic repulsion of all these positive particles in the same place. Some very powerful force, stronger than electromagnetism, must be holding them together. Though strong, it must also operate over a very small space, in fact no more than the diameter of the nucleus. Experiments eventually showed that the strong nuclear force is about a hundred times as strong as the electromagnetic force, but unlike that force (or gravity) it does not follow an inverse square law – it has simply no effect outside the nucleus, giving it a range of $10^{-13}$ centimetres, very small indeed. This, then, is the reason why uranium with 92 protons is the last naturally occurring element (and an unstable one at that), and the why the larger artificial elements created beyond it are also unstable. With that many protons, the electromagnetic repulsion has become strong enough to overcome the strong nuclear force.

When Chadwick identified the neutron in 1932, it turned out not to be the last word in subatomic particles – far from it. Some puzzling features of radioactive beta decay, which gives off a stream of energetic electrons, had lead the Austrian physicist Wolfgang Pauli to suggest in 1930 that there must be another type of particle involved. It emerged that in beta decay, the mass-energy of a neutron is converted into a proton and an electron, which is ejected. The fact that there is another proton changes the type of the element. However the equations did not quite add up, and it was left to the Italian Enrico Fermi (1901-1954) to propose a brand new particle, the neutrino, which as its name suggests, is electrically neutral, very small indeed, and easily able to pass through lead. He showed that the production of a neutrino is triggered by a weak and short-range force, causing a neutron to decay as described. This new field of force soon came to be called the weak nuclear force. This meant that there were now four known forces in the universe – gravity, electromagnetism, and the strong and the weak nuclear forces.

Thousands of scientific man-years have been expended ever since the 1930s in trying to find the equations to unify these forces, which, it is suspected, are different manifestations of some underlying principle. There has at least been some success in bringing together the weak nuclear force and electromagnetism into one model, known as the electroweak interaction. The wider field is known as quantum electrodynamics, which describes how light and matter interact. The best-known proponent of the "QED" and its remarkably accurate predictions is the American physicist Richard Feynman (1918-1988), but despite his attempts to popularize his theory, it really lies beyond the scope of popular science. The attempts at integrating the four known forces go on, but the strong nuclear force in particular has proved a stumbling block, as it is evidently much more complicated than the electric force holding the electrons to the atoms, or gravity.

Since they were identified theoretically, there have been many attempts to capture or count neutrinos, thought to pour out of the Sun in vast quantities. This work involves the construction of large underground tanks containing heavy water – that is, water with an abundance of deuterium (heavy hydrogen – that is, hydrogen with a proton and a neutron in its nucleus; the more common isotope has no neutron). These tanks are placed in old mines, far away from daylight where other forms of radiation would interfere with the experiments.

Also ongoing is the search for esoteric particles below the level of the proton and neutron. This type of work is carried on at a large institution which lies mostly underground, CERN (the European Organization for Nuclear Research), in Geneva, Switzerland, home of the Large Hadron Collider (a hadron being a subatomic particle). Here we enter the strange world of quarks, mesons, bosons, muons, baryons and the like (in fact over 150 subatomic particles have now been identified, most of them existing for the merest split-nanosecond). They are given the generic name "quark", and are classified in six groups – up, down, strange, charm, top and bottom, which on the face of it sounds a funny way to classify anything. The idea is that the quarks are the basic building blocks of matter, and are held together by gluons to form protons and neutrons. Meanwhile electrons and neutrinos are thought to be made of leptons. Quarks and leptons together are called fermions, named after Fermi, discoverer of the neutrino. Bosons, on the other hand, are particles which produce and carry forces, amongst which are gluons and also photons. They are named after the Indian physicist, S N Bose. Of course the most famous of them all is the semi-mythical Higg's boson, named after an English scientist called Peter

Higgs (1929-) who says its existence is required as a way of imparting mass to a particle.

Somewhat like Ptolemy's view of the universe with the Earth at its centre, with all its epicycles and equants, it certainly looks messy. The physicist Leon Lederman put his finger on it when he said "We don't really see the Creator twiddling twenty nobs to set twenty parameters to create the universe as we know it......There is a deep feeling that the picture is not beautiful" (quoted in Bill Bryson). Also, the model as it stands is incomplete, because so far it has no place for gravity, which of course operates at the very largest of scales, and is effectively non-existent at the subatomic level. Nor can it account for mass – rather a fundamental problem – without the notional Higg's boson, which may yet prove to be another version of Einstein's cosmological constant.

This massively expensive research – "Big Science" – has yet to yield much in terms of practical, usable physics. Even the announcement that the Higg's boson had been detected in 2012 was shrouded in doubts about what had really been found. It seems far too early to write the history of this work-in-progress.

# Chapter 13 – The Development of Genetics

By Mendel's time, two things had long been known to biologists – first, that the cell is the fundamental unit of life, and second, that most plant and animal cells contains a nucleus, which must clearly have an important function, as yet undefined. (Red blood cells do not have a nucleus.) In 1858, just before the publication of *Origin of the Species* and whilst Mendel was tending his garden, a German professor called Rudolf Virchow stated his belief that every cell is derived from another cell (which is why the origin of life remains such a mystery). Later scientists working with transparent sea urchins in the 1870s observed that when a sperm penetrates into an egg, the two nuclei fuse to form one, clearly bringing together the heredity of both parents.

In 1879 yet another German, Walther Fleming, identified chromosomes within the nucleus, thread-like micro-organs which can be picked out by staining them with dyes. It was soon observed that these chromosomes replicate when any cell divides, so that both cells have a copy. August Weismann (1834-1914) saw that it must be the chromosomes which carry heredity, but he was more specific than that: it must involve the physical passage of a definite chemical with its own molecular composition, which he called chromatin. He also made the very important distinction between the two fundamental types of cell division. He saw that during regular cell division, all the chromosomes are duplicated before the cell divides, so that each daughter cell has its own copy. However, during the kind of cell division which produces egg or sperm cells, each only has half the full complement of chromosomes. The full complement is only restored when a sperm cell penetrates an egg, as had been observed with the sea urchins So the rest of the body cells are not involved in this process, as Darwin had mistakenly thought. It was also going to be difficult to sustain Lamarck's point of view that learned experiences of the parents can affect heredity. After all, with mammals for example, half the heredity comes from the egg, which is born with the female! The male cells are created within the testes (so are not born with the male), but as Virchow had shown, they too must come from the division of other cells. This left no way in for parental experience, since all the parent's cells come originally from its own parents, starting out from a single fertilised egg.

The Dutchman Hugo de Vries (1848-1935), in as book published in 1889, suggested that heredity must be defined in a number of distinct units. He called these "pangens", a term which later changed to "genes". In the subsequent decade he carried out a large series of plant trials, very much in the manner of Mendel, and came to the same conclusions Mendel had reached long before him. As he was getting ready to publish, in 1899, he surveyed the literature on the subject, only to find – to his mortification – Mendel's long-forgotten papers. Moreover, he was not alone in this experience. Two other researchers, one in Germany and another in Austria shared it with him! This time, the world was clearly ready for Mendelian heredity.

This work was taken up in the next century by the Cambridge scientist William Bateson (1861-1926). H did much to establish genetics as a respectable university science rather than an esoteric form of gardening, and indeed coined the term "genetics" itself. At first he obtained little financial support, and had to rely on the under-utilised labour of the students at a women's college in Cambridge, Newnham. Bateson was able to confirm and extend the findings of Mendel and De Vries, noting, for example, that a pair of black chickens would produce one pure white chicken for every three black ones, just like Mendel's peas. He also conducted a large-scale series of horticultural experiments between 1920 and 1910. These gave rise to one big problem, however, the snapdragon, which seemed able to produce flowers of new colours from nowhere. Painstaking research over a number of growing seasons by one of his pupils demonstrated that this came about because the colour came not from one simple gene, but from a combination of them.

*******************************************

Other scientists approached the study of heredity from quite a different angle to Mendel. One of these was the Swiss biochemist Friedrich Miescher (1844-95), who worked in Germany as well as in Switzerland, and who was to die at the early age of 51 before the true significance of his findings became clear. He set out to find out how cells work in the human body, starting out with white blood cells, which are called leucocytes (a term which comes from the Greek "white cell"). He obtained free cells from bandages from a nearby hospital! He expected to find that the cells were full of proteins, but he found that the nucleus itself contained another substance which had a different chemical structure, for one thing because unlike any protein it contained a significant amount of phosphorus. He called this substance nuclein.

By 1869 Miescher was able to confirm the existence of this previously unknown hydrocarbon, and also that it was found in the cells of yeast as well as other body cells (such as liver cells). He then went on to study the sperm cells of salmon, useful from his point of view because these cells are almost all nucleus. Their only purpose is to pass on a genetic message, so they do not need the full paraphernalia of other cells, which are normally mini-biochemical factories of great complexity. He found lots more nuclein here, consisting of large molecules containing several acidic groups. By 1889 others were looking at this fascinating substance, now known as nucleic acid. Miescher had been on the way to the discovery of the chemistry of inheritance, what we now call DNA, but he died in 1895. However by that time the cell biologists had seen that nucleic acids and chromatin as described by Weismann are one and the same and were likely to be the mechanism of inheritance.

It soon became clear that the basic building block of nucleic acid is a sugar, ribose, attached to a structure containing phosphorus. The final part of this large molecule is its "base", and this comes in one of five flavours – guanine, adenine, cytosine, thymine (not thiamine, which is a vitamin) and uracil, usually abbreviated to G, A, C, T and U. The ribose pentagon at the centre of the molecule caused the whole structure to be named ribonucleic acid (RNA). This particular hydrocarbon in fact only contains four of the bases, G, A, C and U.

It was not until the 1929 that a second and very similar molecule was identified and given the distinctly unwieldy name of deoxyribonucleic acid, unsurprisingly shortened to DNA. In structure this is ribonucleic acid with one less oxygen atom (hence "deoxy-"). This also contains four bases, this time G, A, C and T. It was first identified by an American biochemist called Phoebus Levene (1869-1940). He was born a Jew in Russia, but emigrated to the USA in 1891 after a pogrom at home. Apparently his unusual first name arose from a linguistic confusion as he should have anglicized his Russian name Feodor as Theodore!

Levene was destined to lead the search for the means of inheritance down a blind alley. He was a founding member of the Rockefeller Institute in 1905 and had a very high standing in the academic world.

His collected relatively large quantities of nucleic acid and found that it contained almost equal quantities of G, A, C and U – the constituents of RNA. In fact it seemed probable that each molecule of RNA contained just one of these four bases. This idea became known as the tetranucleotide hypothesis (tetra being the Greek for four). His conclusion was that this relatively straightforward substance must just offer a supporting structure to the real stuff of inheritance, which was expected to be a protein. It was fully realised that the instructions for inheritance must be very complex, and RNA just seemed too simple. Proteins, on the contrary, are extremely complicated. Their recipes contain many amino acids, carefully linked and folded in very complex ways – so it seemed that a protein must be what Levene was looking for. The mistake he made was understandable, though his hypothesis could have been tested!

The next clue in the trail leading to the discovery of the mechanism of heredity came from an unexpected place, not a university but a Ministry of Health laboratory in London. Here in 1928 a microbiologist called Fred Griffiths was studying the bacteria thought to cause pneumonia, known as pneumococci, and he found that there were two types, deadly and harmless. He subjected the dangerous type to heat treatment and then injected it into his mice, which suffered no ill effects. However if he mixed the dangerous (but dead) bacteria cells with the harmless ones, the result once again was lethal. Something was able to pass from the dead bacterial cells into the live ones and change their genetic structure, but nobody knew what it was. Griffiths himself reported his findings but never got to the bottom of the problem, and was in fact killed in the London blitz of 1941.

(Note this first usage of the term *microbiologist* in this book. According to Steve Jones, a molecular biologist – presumably something similar – is just an anatomist writ small, plus an enormous research grant. Things have obviously changed since the days of Fred Griffiths.)

The baton was picked up by another pneumonia scientist, Oswald Avery (1877-1955) of the Rockefeller Institute in New York. He took Griffith's dead cells and subjected them to a process of freezing and thawing to break them open, then spun them in a centrifuge to separate the liquids from the solids. The dangerous agent was found in the liquid fraction. A long period of research with two assistants, MacLeod and McCarty, finally identified this agent as DNA, and this result was published in a paper in 1944. This was read by an Austrian scientist, Erwin Chargaff (1905-2002), who had settled in the USA at Colombia

University in 1935. He began a detailed study of DNA, finding that this same substance was present in each species he examined. He noted that guanine and adenine belong to one chemical group (the purines), while cytosine and thymine belong to another (the pyrimidines). He published a paper in 1950 which showed that the amount of G + A is always equal to the amount of C + T; that the amount of A is always equal to the amount of T; and that the amount of G is always equal to the amount of C. These findings, known as the Chargaff rules, turned out to be critical to the way that DNA works.

Meanwhile direct studies of heredity itself had been taking place. The leader in the field was an American zoologist called Thomas Morgan (1866-1945) of Columbia University. He was sceptical of the newly re-emerged Mendelian "factors" (genes), considering them at best a special case applying only to a few features. For his own research he selected the most famous fly in all science, *Drosophila melanogaster* ("black-bellied lover of the dew"), since his time the "workhorse" of genetics because it breeds so fast and because when all is said and done, nobody much cares about flies. It also has the advantage that it possesses only four pairs of chromosomes.

After two years of failures, Morgan started getting results when studying the sex chromosomes. Because these are so different between male and female, they have been a fruitful area of research ever since. Females always carry two X chromosomes (XX), and males an X and a Y (XY) – the combination YY is simply not possible as the mother has no Y, and would not in any case be viable because the Y chromosome is very small, being mainly just a sex determinant. All creatures need at least one X, which holds other genes. Any new creature must inherit an X from its mother. It may inherit either an X or a Y from its father. If it inherits a Y, it is a male.

Morgan's fruit flies turned out to have two eye colourings, a common red and a rare white. He found that when a male with white eyes is bred with a female with red eyes, all the offspring have red eyes – so from Mendel's laws, the gene for red must be dominant, and the gene for white recessive. However when a male with red eyes is crossed with a female with white eyes, all the male offspring have white eyes, and all the female offspring have red eyes. What did this mean?

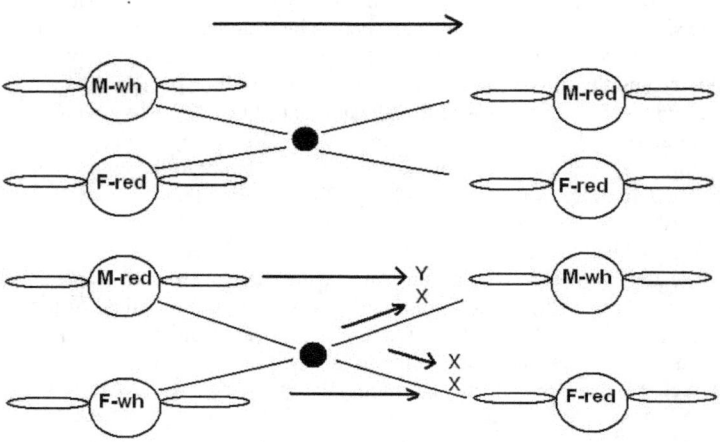

With male dominant red and female recessive white, the white in the next generation can only come from the mother, and only goes to the sons, because the daughters must get a red from their fathers

Male offspring always look like their mother. Morgan concluded that the gene for eye colour must be resident on the X chromosome. White eyes would only show up in the next generation if the mother has them on both her X chromosomes. If the father has red eyes and the mother white (the second case, above), the female offspring MUST inherit this single X chromosome from the father (as he only has one X) and so must have red eyes. The male offspring DO NOT INHERIT an X from their father at all. They inherit a Y. Therefore their eye colour comes from their mother, and she MUST have two genes for white, as red is dominant. Moreover the gene for sex must be in the same place as the gene for eye colour, a phenomenon known as sex linkage. Get the idea?

Sex linkage was a development of Mendel's rules, because he had identified that different characteristics are independent of one another, and in this case clearly they were not. So, having got this far by 1910, Morgan continued his researches. Next he found that the process which makes a sperm or an egg, each of which has only one set of chromosomes instead of a pair, selects genes in blocks from one chromosome or the other of the parent. This process is called recombination, a reshuffling of the pack. An individual sperm, for example, will obtain all its heredity from the male who makes it, but it only has room for half of what he has, because it needs to fuse with an egg which has half whatever the mother contributes. This of course is the reason why children can be so different from either parent. They get 100% of their genetic material from their parents, 50% from each, but they may in fact get the recessive 50% which neither parent is using very much.

So a splicing process takes place, which means that gene A which lies physically close to gene B on the chromosome is much more likely to be transferred with B than with C, if C lies far away in the chromosome. Morgan set all this out in two books, *The Mechanism of Mendelian Heredity* (1915) and *The Theory of the Gene* (1926).

More studies conducted since Morgan's time have shown the disadvantage of being a male with only one X chromosome, as this can carry many inherited diseases. In a girl an X chromosome which is in some way abnormal may not find expression because of the presence of the other X chromosome, but in a boy there is no alternative. So sex-linked abnormalities are much more common amongst boys; hereditary muscular dystrophy is one of them.

*******************************

A scientist in a related field was Hans Krebs (1900-1981). Another Jewish refugee from Germany, he did his most important work at the University of Sheffield before the Second World War. He identified two essential metabolic processes, the urea cycle and the citric acid cycle. The latter, called the Krebs cycle, is the sequence of chemical reactions which take place in individual cells to produced energy. In principal, this involves the oxidation of carbohydrates, fats and proteins to produce chemical energy in the form of ATP (adenosine triphosphate), with carbon dioxide as a waste product.

*******************************

The X-ray spectrometer developed by William and Lawrence Bragg came to be used to probe the very tissues of the body, much of this work being undertaken at Caltech in California under Linus Pauling (1901-94). Pauling, who came from Portland, Oregon, and descended from German farmers, was to be one of the most influential scientists of the mid-century. He was able to demonstrate how the hydrogen bond holds the structure of proteins together. At the beginning of the 1950s this work had extended to fibrous tissues including hair, fingernails and wool (all formed from proteins, the first two from keratin). In fact these are made of polypeptide chains, which are strings of amino acids with the water squeezed out of them and which form the backbone of any protein. Pauling's team at Caltech found that fibrous proteins have a structure where the polypeptide chains are wound around one another in a spiral, like a twisted rope, and held together with hydrogen bonds. They called this an alpha-helix structure. During 1951 they were able to issue a number of papers giving details of the structure of hair, feathers, muscles and so on. This work was considered a triumph and was indeed a great coup for Pauling over his rival, Lawrence Bragg, who had invented the X-ray spectrometer with his father and who was now in charge of the Cavendish Laboratory in Cambridge.

(Incidentally Pauling is the only person ever to receive two unshared Nobel prizes in different categories. He was awarded the prize for chemistry in 1954 for his work on chemical bonds in the 1930s and for his work on proteins leading to the 1951 papers. He was also awarded the prize for peace in 1962 for his part in obtaining an international ban on the testing of nuclear weapons above ground.)

It was now clear that Pauling would try to crack the structure of DNA, not a protein but clearly by now central to the heredity puzzle. However, there were also two teams in England which were capable of taking on this project, one at the Cavendish in Cambridge and the other at King's College, London under Maurice Wilkins (1916-2004). In these straitened post-war times there was only enough funding for one team to research this single project, and by gentleman's agreement, that was the King's team. Into this cosy arrangement barged a young postdoctoral student from the USA, James Watson (1928-), who had made up his own mind to crack the DNA code, and who was not inclined to show much respect for any gentleman's agreement. He found himself sharing a room with an older PhD student, Francis Crick (1916-2004), who was working on X-ray studies of proteins. Crick had read a little book written by the famous physicist Erwin Schrödinger, of cat and quantum wave equation fame, called *What is Life?* In the book, Schrödinger speculated that the genetic code might be written out in a series of letters from a short alphabet, rather like the Morse code which consists only of dots and dashes, yet which can spell any word in the English language. Despite twice being told to keep off DNA by Lawrence Bragg, Crick could not help himself, and he and Watson pursued the matter unofficially.

Crucial in this matter were the new X-Ray diffraction photographs being produced under Wilkins at King's by Rosalind Franklin (1920-1958) – in California, Pauling only had pre-war photos. Into the picture now stepped another scientist, John Griffiths, son of Fred of the pneumonia studies, and now at Cambridge. When Crick became stuck with his model Griffiths suggested that adenine and thymine would fit together physically, as would guanine and cytosine, the two sets being linked by hydrogen bonds. In the middle of all this, in July 1952, in walked Erwin Chargaff, who explained his rules, previously unknown to Crick. These state that there are always equal amounts of A + G, and also of C + T; and that the amount of A equals the amount of T, and G of C. This suggested that A must be paired with T, and G with C, as the Griffiths model had indicated.

In January of 1953 the team at the Cavendish heard that Pauling had solved the puzzle of DNA, and received an advance copy of his paper from his son Peter Pauling, who was also at Cambridge. Initially downcast, they found that the American model, which plumped for a triple helix, did not match the X-ray diffraction photos coming from Franklin. It must be wrong. Watson took the paper to Wilkins in London, and Wilkins produced one of Franklin's clearest photographs which showed to the initiated that beyond a doubt it was a helix, but only a double helix. Franklin herself had put two and two together and had drafted a paper of her own. In a burst of feverish activity, Crick and Watson put together their paper, only 900 words long, which included a diagram of the double helix drawn by Crick's wife. This paper appeared in *Nature* on 25 April 1953, stressing its agreement with the Chargaff rules. With it was another paper from Wilkins and his colleagues drawing the same conclusions from the X-Ray diffraction photographs. A third paper from Franklin and her assistant Raymond Gosling gave the X-ray evidence in convincing detail. Bragg and his Cavendish team had trumped Pauling at last. The Nobel prize for Physiology or Medicine went to Crick, Watson and Wilkins in 1962. Franklin, who may also have received a share of the prize, had died in 1958, and Nobel prizes can only be awarded to the living.

There has always been a certain amount of controversy about this discovery, one of the great scientific coups of the twentieth century, because of the role of Rosalind Franklin. It is claimed, notably by feminists, that she was deliberately kept out of the picture by Wilkins, who didn't much care for women in the laboratory. However, her citation on the paper containing most of the solid evidence would certainly have carried Wilkins' name as well, had it appeared alone, as she was not working on her own initiative but following his instructions. James Watson put it on paper that Franklin was uncooperative, unreasonable, secretive and (worst of all) willfully unsexy. She herself posted notices to the effect that the double helix theory was wrong.

It seems that she was satisfied at the time, as indeed her paper was included in *Nature*. That same year she moved on to more agreeable employment at another London college, Birkbeck, but died of ovarian cancer five years later at the age of only 37. It seems that, like Pierre Curie, she fell victim to her science, having over-exposed her body to X-rays, and is said rarely to have worn the protective lead apron which was available to her.

It soon became clear that the Chargaff rules were right, as it was found that in the double helix, an A on one helix always lines up with a T on the other, as does G with C. The two strands are like mirror images or complements of each other. When the strands unwind to make a new cell, all the copying process has to do when it finds an A is to make a T, and so on.

Crick continued his work on the double helix, whereas Watson made no other significant contributions. It emerged during the 1950s and 1960s that the genetic code is written in triplets called codons, for example CTA, each of which represents an amino acid. A codon is literally a code for an amino acid, not a recipe, but the codons are strung together in genes to make a recipe for a protein. There are 20 amino acids in the body, which can be mixed in thousands of combinations to make individual proteins. The genes themselves vary greatly in length, from about 500 letters to more than two million. The whole system is uncannily close to the binary system used in computers, with four letters instead of two – it is a digital system, invented by nature.

In the course of his work, Crick expounded an idea which is so clever that it has been called the greatest wrong theory in history. This was his "comma-free" code. He deduced that the codes for individual amino acids must each consist of three letters, but there are 64 possible combinations, which he whittled down to the 20 needed for the amino acids by a process which eliminated 3-letter codes which could be misread if the reading started in the wrong place. Experimental evidence was eventually to show, however, that the real scheme uses different 3-letter codes for the same amino acid, so it contains redundancy and is less elegant than Crick, but more tolerant of errors.

In the process of protein manufacture, part of the double helix is unzipped and the required code is read off and copied into "messenger"

RNA, which is has the same structure as DNA except that every T is replaced by U (uracil). The codons are assembled into the RNA and are then passed into a ribosome, which is the cell's own protein factory. When the job is done, the components of the RNA are recycled.

Mistakes can be made in the copying process. If the error occurs in a single daughter cell, it may have little effect, unless that happens in the very early stages of embryo growth. If however it happens in the rather more complicated splicing process which produces a sex cell, then it will affect every cell in the body of the new individual. This provides a mechanism for Darwin's heredity. Mutations at this level can produced individuals which are better suited to their environment, though usually the individual will be less well suited. The body has its own checking processes, and during the embryo phase, deleterious changes can result in the spontaneous abortion of the embryo.

It has become apparent that Darwin was also quite right to claim human descent from the apes. There are different measures, but the amount of human DNA shared with chimpanzees and gorillas is generally put at around 98%. However as there are clearly quite important differences between man and the apes, it is apparent that there is more to it than that. A lot of the difference is thought to revolve around the controlling switches which lie embedded in the DNA, which tell a process when to start, and when to stop. The best-known of these are the hox genes which control the growth of an embryo and tell one end of it to be a head, and the other the feet. However it is a fact that the vast majority of human DNA does not appear to code for anything – the so-called junk DNA. Recently, however, it has been found that this "junk" is the home of the numerous control switches, possibly millions of them. Also, the human DNA is contained in 46 chromosomes in each cell (two pairs of 23), whereas chimpanzees and gorillas each have 48 chromosomes – amongst other things, this would form a complete barrier to interbreeding.

It also emerged, to general surprise, that humans have only about 30,000 genes, only twice as many as the average fly, and no more than some common plants! Again, it has become clear, it's not what you've got, it's what you do with it. Moreover it has been found that there are certain basic housekeeping functions of the cell which must be performed by all living things, which therefore share virtually identical genes for these purposes. In fact about half the biochemical processes taking place inside a banana also take place inside people.

In fact DNA can be quite an unreliable guide to outer form, and indeed some creatures have a life-cycle which exhibits entirely different forms (caterpillar and butterfly for example) from the SAME DNA. One example of this is the population of cichlid fish in the Great Lakes of East Africa. There are hundreds of species filling every ecological niche – grazers, predators, parasites – all different shapes and sizes; but all sharing very similar DNA.

So it is that the genes, made of DNA, act as a library of instructions for the body. What we inherit from our parents is really a set of recipes and a set of control switches to organize the recipes. The executive functions of the body are performed by the proteins, not the genes, and the proteins are made of amino acids, though folded and shaped in many wondrous ways. These are two completely different biological systems, and the link between them is the RNA.

The Human Genome Project completed the mapping of each of the 3 billion letters in human DNA in 2001, and the DNA of many plants and animals has now also been unraveled. This work, painfully slow at first, was speeded up by advances in technology. One of these is the polymerase chain reaction, which literally sets up a chain reaction by feeding the DNA with a supply of chemicals and allowing its natural copying ability to make many new copies. It has become abundantly clear that all of life is one, from the filaments in the ponds of Yellowstone to the tube worms of the black smokers at the bottom of the oceans, even to viruses which are not wholly alive. There is only one code, one DNA, one RNA, though nobody knows which came first – the DNA or the RNA. Watch this space.

The very idea of genes and DNA has stimulated some interesting literature, perhaps none more so than *The Selfish Gene* (1976) by Richard Dawkins. His idea is that people and other creatures are merely mindless replicators, working for the gene, which ruthlessly exploits its carrier in order to reproduce itself. In some cases, as with Atlantic salmon, the very body of the creature is used up to make sex cells on its final journey, and is cast away immediately after spawning. Altruistic behavior is likely to be found amongst blood relatives who share genes. It is an extreme case, as altruistic behavior has been noted not only amongst unrelated people, but even amongst unrelated species. Dawkins later said he should have called his book *The Immortal Gene*, but of course, that might not have sold so well.

\*\*\*\*\*\*\*\*\*\*\*\*\*\*\*\*\*\*\*\*\*\*\*\*\*\*\*\*\*\*\*\*\*\*\*\*

X-ray crystallography was to play a part in another famous discovery of the twentieth century, penicillin. This story is one of the least well understood matters in science as far as the general public is concerned, because virtually all of the credit is generally ascribed to one man, Alexander Fleming, when in fact all he did was to set the ball rolling. It would be ten years before anyone picked up that ball.

Fleming (1881-1955) was the son of a farmer from Ayrshire in Scotland. He qualified as a doctor in 1906, but by 1928, after a period of research and service in the First World War, he was the Professor of Bacteriology at St Mary's Hospital in London. It was in that year that he went on holiday in August, leaving some petri dishes containing his cultures lying around in his laboratory. When he returned from his vacation he found that one of his cultures had been contaminated by a fungus, and the colony of bacteria adjacent to the fungus had been killed off. He had discovered penicillin by accident. However, experiments showed that it only killed off certain types of bacteria.

Fleming published details of his findings in 1929 in the British *Journal of Experimental Pathology*, but finding himself unable to make much progress, eventually moved on to other research. There were several problems. Firstly, it proved difficult to produce penicillin in meaningful quantities. Secondly, he did not have the chemistry skills he needed to isolate the active ingredient. Thirdly, he suspected that it would not last long enough in the human body to have much effect on infections. That is really Fleming's complete contribution to the development of penicillin. However, there is no doubt that Fleming fully realised the antibiotic properties of his discovery.

The baton was then picked up by the Australian Howard Florey (1898-1968) of Lincoln College and the Radcliffe Infirmary in Oxford. In 1939 war was breaking out. Large numbers of casualties were to be expected, and a cure for bacterial infections was urgently sought. Florey and his team, notably Ernst Chaim, a Jewish refugee from Eastern Europe, and Norman Heatley, thumbed their way through academic journals looking for something which may be of use. They found Fleming's paper and immediately began their own researches. However, they soon bumped into the same problems which Fleming had encountered. Nevertheless by 1941 they had enough of the drug to administer it to a dying patient, though little idea of the dosage. In the event the patient, one Albert Alexander – who had scratched himself on a rose and become infected – died anyway; but he had shown a marked improvement before the minuscule supplies available to treat him ran out.

The biochemistry of penicillin proved very complex, and it became evident that even the Oxford team did not have the resources to produce significant quantities of the drug. Because of wartime imperatives, the matter was passed on to the American government, which accorded it the second highest scientific priority of all, only behind the Manhattan project (to produce the atom bomb). The US government passed on the work to the major American pharmaceutical companies including Merck & Co. A much smaller company, Pfizer, came up with a method of producing the mould in quantity. It was found that cantaloupe melons grown in Peoria, Illinois produced significant quantities of penicillin if allowed to go mouldy, and mass production went ahead using this mould, which was cultured using corn steep liquor. Soon vast numbers of melons were being grown and allowed to rot, but it worked. The big pharmaceutical companies including Glaxo in Britain started mass production, and 2.3 million doses were available for the D-Day invasion of Normandy in June 1944. By now one of the chief virtues of the new drug had become apparent – for most people who take it, there are few side-effects.

Ernst Chaim's chemical structure of penicillin was not finally confirmed until 1945, when Dorothy Hodgkin (1910-1994), also of Oxford, used X-ray crystallography to reveal its secrets. Hodgkin went on to decipher the structure of vitamin $B_{12}$ and also of insulin. She was awarded a Nobel prize for her work on vitamin $B_{12}$. Fleming, Florey and Chaim shared another Nobel prize for their discovery and development of penicillin, though it was Fleming who garnered most of the publicity. Without a doubt, this was the greatest life-saver of the twentieth century, and the greatest advance in medicine since Jenner and his smallpox vaccination, if not of all time.

*******************************************

The development of penicillin was followed only eleven years later by another great breakthrough, a successful vaccine against polio, a disabling or even fatal virus which had only appeared in 1909. This is credited to Jonas Salk (1914-1995), another American of Jewish ancestry, who led the fight against the greatest remaining scourge of twentieth-century childhood. Up until the appearance of the vaccine in 1955, polio was feared by every parent of young children – there were 58,000 cases reported in the United States in the epidemic of 1952, and the epidemics were annual. This was the worst-ever figure – the disease seemed to become more rampant. It was the same story in Britain. Of those affected in the American epidemic of 1952, over 3,000 died, and many more were left with severe disabilities. The most prominent victim of all had been President Franklin D Roosevelt, who spent his working life in a wheelchair, and who set up the fund to search for a vaccine.

Working at the University of Pittsburg, Salk devoted the seven years from 1947 to 1954 developing his vaccine before massive and successful field trials in 1955 (in which 1.8 million schoolchildren were involved). However there was never any issue of a patent for the vaccine. This was really much more typical of the British attitude towards medical research, where the patenting of such discoveries was regarded as virtually immoral; neither aspirin nor penicillin had been patented. Salk's program had been publicly funded, the vaccine was for the public good and that, as far as Salk was concerned, was the end of the story.

However, Salk was not the only scientist working in this field. A decade previously, another American, Hilary Koprowski, had developed a vaccine which had to be withdrawn when it was found that it could actually spread the disease. Another American contemporary, Albert Sarin, also worked on a different and ultimately successful version of a vaccine, which, although appearing slightly later than Salk's, was more acceptable as it could be taken on a lump of sugar. This was the vaccine adopted in the United Kingdom in 1959.

\*\*\*\*\*\*\*\*\*\*\*\*\*\*\*\*\*\*\*\*\*\*\*\*\*\*\*\*\*\*\*\*\*\*\*\*\*\*\*\*

Life is thought to have emerged in single-celled forms fairly early in the history of the earth, probably by 4 billion years ago. Charles Darwin famously imagined that this event took place "...in some warm little pond, with all sorts of ammonia and phosphoric salts – light, heat and electricity present..." However, even a single-celled organism has proved to be so immensely complex that all attempts to replicate the warm little pond, and to create life from scratch, have so far failed. In 1952 an experiment conducted by Stanley Miller and Harold Urey at the University of Chicago seemed to be getting close. An atmosphere thought to resemble that of the early earth was created, containing ammonia, methane and steam. When electrical charges (test-tube lightning) were passed through it, a rich brew of organic chemicals resulted, including amino acids. These are the building blocks of proteins, which in turn are the building blocks of life. This naturally caused great excitement. Then in 1961 the Catalonian Juan Oro concocted another brew and managed to create the chemical adenine, which is one of the four "nucleic acids" (found in the nucleus) which are the building blocks of DNA, and which is also a major constituent of ATP – which provides cells with their energy.

It has subsequently been shown to be relatively easy to create these organic chemicals from a recipe in a laboratory, but quite impossible – as yet – to create proteins or DNA from them. In theory all that is needed is a chemical which can replicate itself, and a sac or cell to contain it, and then this simple construction can start to evolve. A Danish scientist called Steen Rasmussen is working on a project to make just such a cell, but his sac is made of fatty acids – and these themselves are complex carbohydrates.

For a very long period, life on earth is thought to have consisted of nothing but single-celled bacteria and another group, found in modern forms for example as red filaments in hot, chemical volcanic pools in the Yellowstone National Park, called the archaea. These strange and primitive forms of life were only revealed as a separate kingdom in 1977 by the American microbiologist Carl Woese (1928-2012), who used RNA analysis to show that they are quite separate from the bacteria in their genetic structure. The most primitive forms of life known today are found in an unlikely place – around suboceanic black smokers. They are hyperthermophiles, lovers of extreme heat, surviving on hydrogen and sulphur in anaerobic environments (that is, without free oxygen). At these depths, they have no need of either oxygen or sunlight, and would have survived any surface catastrophe, including meteor strikes and snowball earth. It looks increasingly likely that these bacteria and archaea are the foundation of all other life on earth.

If it is quite clear that all known forms of life have a single common origin, it is equally clear that even the simplest life forms are very complex. They must have evolved to replace all traces of the primitive forms from which they evolved, kicking the ladder from beneath them for the emergence of any other form of life.

The early forms of life (including cyanobacteria, commonly known as blue-green algae, which are still with us) did not possess a cell nucleus – the prokaryotes. The first organisms to contain this structure are called the eukaryotes, and they too have lived on earth for a long time, 1.5 billion years or more. The eukaryotes greatly increased the sophistication of life by including within their cells not only nuclei, but also other structures known as organelles. The two most important of these are mitochondria, of which more below, and chloroplasts, used for photosynthesis. It is thought that these first existed as independent bacteria, but became incorporated into the eukaryotes in a symbiotic relationship. This idea was first proposed by the American biologist Lynn Margulis (1938-2011) in 1966 and later developed in her 1970 book, *Origin of the Eukaryotic Cells*.

Archaea territory – hot pools and travertine formation at the Mammoth Hot Springs in the **Yellowstone National Park**

Eukaryotes may be more advanced than prokaryotes, but they are still only single-celled creatures. It was the development of multi-celled organisms which led to higher forms of life, where groups of cells could specialize to form skin, muscles and so forth. No trace of such creatures is present in the geological record for as much as 900 million years after the proposed appearance of the eukaryotes.

Eukaryotes must have evolved from combinations of prokaryotes, but this is thought to have been a one-off development, because all eukaryotes – including all plant and animal forms – share the same basic structure. Some scientists think the event so rare that it may never have occurred at all on other planets in the universe which possess prokaryote life forms. It may mean that man is the only creature which can send radio beams into space.

\*\*\*\*\*\*\*\*\*\*\*\*\*\*\*\*\*\*\*\*\*\*\*\*\*\*\*\*\*\*\*\*\*\*\*\*\*\*

During the 1950s the details of the process of photosynthesis were further defined. In fact this work was only made possible by the availability of a radioactive isotope of carbon, that is carbon-14, which enabled the progress of carbon molecules to be tracked. It became known as the Calvin-Benson cycle, after two scientists working at the University of California at Berkeley. In fact, the subordinate Benson made the important breakthroughs, but was later pushed out by the head of the department, Calvin, who alone received the Nobel prize in 1961. The cycle describes the complex series of reactions which take place during photosynthesis within the chloroplasts. It is independent of light, because the energy required from sunlight has already been extracted by this stage. The end product is sugar, but this is not produced in a simple reaction – in fact the details of what exactly happens are still unclear.

Very significant developments were also taking place in the field of applied botany, where after decades of research, especially in Russia and the United States, useful results were finally obtained. The man who created the first successful hybrid crop was the American Norman Borlaug (1914-2009). Working in Mexico, he established a variety of wheat which gave good yields and offered resistance to disease, but which blew over in strong winds. The American occupation of Japan after 1945 resulted, amongst other things, in the discovery of a natural mutation of wheat called Norin 10, which grew on the mountains of northern Japan. This strain had developed a short stem which gave it greater hardiness. Borlaug cross-bred his own strain with Norin 10 and produced the first of many short-grained cereals which have since spread around the world. After the introduction of his wheat into the Indian subcontinent in the 1960s, major famines ceased to occur there – as they had done right up to that time – so Borlaug became known and the "man who saved a billion lives."

Also working in this period was another American, Barbara McClintock (1902-1992), who completed her most famous experiments on maize (corn) plants. In commercial varieties, this always has the familiar whitish or yellow seeds, but the original varieties actually had red seeds. In her thousands of experiments, McClintock observed that the "mutation" which caused the yellow colour tended to relapse, so that the plants she bred contained a variety of colours, white, yellow, blue and red, or individual seeds of mixed colour, all on a single cob. However, mutations were meant to be permanent. She eventually concluded that the seed colour was the result not just of a single gene, but of switches within the genome which control the colour genes. Genes for different colours must be present in a single genome.

\*\*\*\*\*\*\*\*\*\*\*\*\*\*\*\*\*\*\*\*\*\*\*\*\*\*\*\*\*\*\*\*\*\*\*\*\*\*\*\*\*

During the course of the 1960s it was discovered that there is another type of DNA, quite apart from nucleic DNA – mitochondrial DNA (mtDNA). This is stored in the mitochondria; thousands of these mini-machines exist within each cell, outside of the nucleus. The purpose of the mitochondria is to burn oxygen to generate energy for the cell. The output from this process is a chemical called ATP which fuels other processes taking place inside the cell. It is thought that the mitochondria originated as separate bacteria, with their own apparatus for inheritance.

There is not a lot of DNA involved – in fact it codes for just 37 genes. However this DNA has its own pattern of inheritance as it is passed only down the female line, without change, so that everyone inherits their mitochondrial DNA from their mother. Nucleic DNA is reshuffled at every generation in the recombination process which creates every sex cell, so it is not as much use in tracing patterns of inheritance as might be thought. Because mitochondrial DNA never changes, except by mutation – and it does mutate much more quickly than nucleic DNA, at the rate of about one change per 3,500 years –

then it has proved ideal for the study of inheritance of the female line (in fact it is the only practical means of doing this). It is no use for the study of the male line, as it leads straight back to any man's mother, but there is something else which can be used for this – the Y-chromosome, which every man inherits from his father.

When scientists started to study the pattern of geographical variation in mtDNA and its rate of mutation, and to add up the statistical probabilities, they came to the conclusion that all the people living on the Earth today must inherit their mtDNA from a single woman. That woman lived in Africa between 150,000 and 200,000 years ago. This idea was explained in an article in *Nature* by Rebecca Cann and her colleagues in 1987. This "African Eve" would have been living in a community with other women, who would have produced their own children, but these lines would eventually die out as the possible source of all mtDNA because each would eventually produce only boys, whose children would inherit their mtDNA from their mother, not their father. The descendants of these other women are certainly still around today, but their mtDNA can ultimately be traced back to Eve. So Eve is the most recent common ancestor on the matrilineal line. Despite widespread amazement when the idea was first published, it is widely accepted today.

If there was a mitochondrial Eve, there was also a Y-chromosome Adam! Similar studies trace the line back to the most recent common ancestor on the male side. Also an African, he lived about 80,000 years ago, so unfortunately he never met Eve! Y-chromosomes, like all other human chromosomes, show much more genetic variety within Africa than anywhere else. The Y-chromosomes of the rest of the world represent nothing more than a small subset of what is found in Africa. This evidence suggests that both Adam and Eve were Africans. Adam was much more recent that Eve because of the different ways men and women reproduce. Woman tend to have a small number of surviving children, but most of them have them. On the male side, from the evidence of primitive societies, large numbers of males are simply excluded from the party. There have been studies showing tribes of a hundred people where all but a few children belong to four men. The big chief may monopolise the fertile women – Genghis Khan is thought to have 16 million descendants! This practice greatly shortens the distance back to the last common male ancestor.

In recent years there have been many publications based on the genetic mapping, showing the movement of peoples by differences in their genes. This has shown up some remarkable facts. For example, in

*The Language of the Genes*, Steve Jones claims that all native North Americans share only four mtDNA lineages, indicating that only small groups of people ever crossed over the Ice Age land bridge where the Bering Straits are now found. Genetic patterns also indicate that Australia was populated by 50,000 or even 60,000 years ago, long before Europe, which was covered in ice at the time. This is despite a sea voyage of 90 kilometers (56 miles) across a deep sea trench which could never have formed land, the presence of which was known to Alfred Wallace (his Wallace Line). Remnants of a "negrito" population still remain on the long road from Africa to Australia, as in the Andaman Islands, Malaya and the Philippines.

# Chapter 14 – Earth Sciences 1890-1970

The Austrian geologist Edward Suess (1831-1914) published a book, *The Face of the Earth (Das Antlitz der Erde)* during the 1890s. In this he proposed the idea of Gondwana, a combination of the land areas of all the southern continents – Africa, South America, India, Australia and Madagascar (and later, Antarctica) – which existed until the Cretaceous period, a hundred million years ago. A body of evidence had built up in support of the idea that the continents as we know them are not a permanent feature of the Earth, and have been in different places in the past. It had long been remarked that the east coast of South America fits remarkably snugly into the "armpit" of West Africa. Again, Madagascar can be made to fit very closely to the coast of East Africa. Matches such as these work even better when the edge of the continental shelf (where the continental crust dips sharply down to the ocean floor), rather than the edge of the current coastline, is used in the fit.

By the time of Suess, it had also become known that there exists a similarity of fossils in South Africa, South America and Australia, notably of *Glossopteris* (a seed fern with a leaf shaped like a tongue, its name based on the Greek word for "tongue"). It was first identified in Permian (299-250 million years ago) sandstones in South Africa, then in conglomerates in India, but it is unknown in the northern hemisphere. This indicated a common history and implied the existence of an early super-continent which Suess called Gondwana. The name is taken from the name of a tribe in India, the Gonds. It means Land of the Gonds so the "land" in the usage Gondwanaland is superfluous.

Another strand of evidence had emerged from the geological field mapping of colonial areas in the later part of the nineteenth century – the presence of glacial tills (indurated boulder clays) from an ancient ice age (in fact Permo-Carboniferous), also in Gondwana. This implied that Gondwana was likely to have been over, or at least near, the south pole, nowhere near most of it today. It was also noted that there is much similarity in the Precambrian (older than 542 million years) sequences of Central Africa, Madagascar, southern India, Brazil and Australia.

Suess, writing before the discovery of radioactivity, also thought that the Earth was cooling and so shrinking. This would necessarily lead to a contraction of the existing crust, which would cause mountains to be thrown up as the segments of it struggled to fit into a smaller space. For the time, this seemed a reasonable explanation, but it is now known that the earth is neither cooling nor contracting.

Following the startling findings of Ernest Rutherford, whose studies of uranium indicated that the Earth could be a great deal older than had previously been thought, new ideas came forward. One man who took a central role in geochronology was the English geologist Arthur Holmes (1890-1965), eventually Professor of Geology and Edinburgh University. In 1913 Holmes published a booklet called *The Age of the Earth*, in which he estimated the age of the earth at 1.6 billion years, at that time far older than any earlier estimates.

In 1915, twenty years after Suess published his great book, there was a further development from the original conception of Gondwana, and this was contained in another book (only 94 pages long in its first edition) called *The Origin of the Continents and Oceans*. A German meteorologist called Alfred Wegener (1880-1930), having studied what evidence there was, proposed the idea of "continental drift" whereby ALL the continents floated apart on a bed of oceanic crust. However there was much uncertainty about the mechanism by which this could take place. (Wegener died in 1930 aged only 50 on a meteorological expedition on the ice of Greenland, his hypothesis still unproven.) However the idea was not without its supporters, one of whom was Arthur Holmes. He championed Wegener at a time when continental drift was deeply unfashionable with his more conservative peers. As for the mechanism by which it could take place, Holmes – presciently, as it turned out – proposed that Earth's mantle contained convection cells that dissipated radioactive heat and moved the crust at the surface.

\*\*\*\*\*\*\*\*\*\*\*\*\*\*\*\*\*\*\*\*\*\*\*\*\*\*\*\*\*\*\*\*\*\*\*\*

It was only towards the end of the First World War that significant progress was made in that most diffuse and troublesome area which had resisted all attempts at classification – the analysis and forecasting of the weather. Inspiration came at last from a place which certainly gets plenty of it, Bergen in Norway, where it is said to rain on 364 days of the year, and don't forget your umbrella on the 365th. Observers at the

Bergen School of Meteorology under Jacob Bjerknes came up with a credible way to analyse the shifting patterns they saw in the sky above them, now known as the low-pressure cyclones of the North Atlantic.

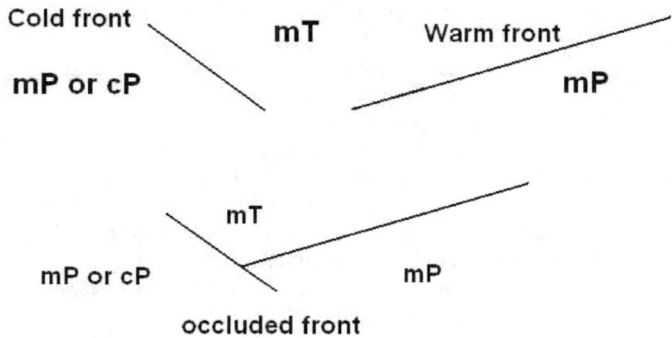

Maritime tropical air gives rise to a warm front ahead of it and a cold front behind it

They envisaged warm, tropical air rising above a sloping wedge of cold polar air as it advances northwards across an area of low pressure. This they called the warm front – in fact as we can see, a front between two quite different types of air. Moisture in the rising warm air then condenses and falls as rain. The warm front is marked on maps as a line with rounded bumps on it. The warm front is often preceded by a mackerel sky of altocumulus cloud.

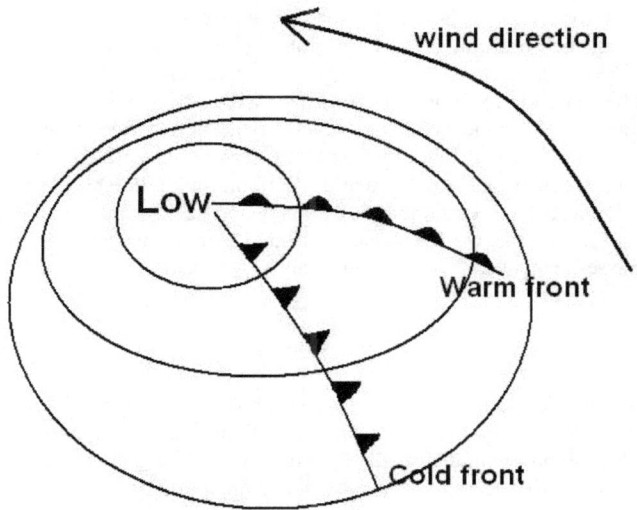

Coming along behind the warm air the Bergen men then detected more cold air, coming from North America (continental) or the Arctic (maritime), heading generally eastwards and pushing up the warm air ahead of it, and so causing more precipitation as the warm, wet air rises. This they called the cold front, and marked it on maps with a line with sharp arrowheads.

In both cases, convergence is occurring, where two air masses come together in one place. Convergence leads to the uplift of one of the masses, low pressure and precipitation. (Divergence is the opposite – sinking, drying air and high pressure.) Both fronts cause a distinct kink in the isobars (lines of equal pressure), felt on the ground as the wind freshens and veers as the front passes.

Where the cold front has caught up with the warm front, an occluded (closed) front forms, where the warm air has been lifted right off the surface and only exists at higher altitudes.

This simple, straightforward picture is often a little more confused on real weather maps, where a single front may be shown as warm, cold and occluded along its length, but the term front – taken from the military fronts of the First World War – remains useful. It retains the idea that fronts are places where air masses of different characteristics meet. These are not confined to the North Atlantic, but are found round the world, as for example in North America where cP (continental Polar) air from the Arctic meets mT (maritime Tropical) air from the Gulf of Mexico, a confrontation which gives the continent much of its weather. The same kind of thing happens over west Africa, where hot Saharan (continental Tropical) air meets the wet maritime tropical air from the Gulf of Guinea. In fact it is perfectly sensible to define only four different types of air in the world: mP (maritime Polar), mT (maritime Tropical), cP (continental Polar) and cT (continental Tropical). Much of the world's weather forms from fronts between these air masses, but not all of it – hurricanes, for example, develop purely within mT air.

\*\*\*\*\*\*\*\*\*\*\*\*\*\*\*\*\*\*\*\*\*\*\*\*\*\*\*\*\*\*\*\*\*\*\*\*

Wegener produced several new and longer editions of his book before his untimely death in 1930, but meanwhile some light was being cast on another geological problem, the cause of the Ice Ages which gripped the Earth from the start of the Quaternary period 2.58 million years ago. Fluctuations in the amount of heat received from the sun known as Milankovitch cycles are nowadays thought to have caused the massive swings in temperature which occurred during the Ice Age itself. Mulitin Milankovitch (1879-1958) was a Serbian civil engineer and mathematician who worked out his theory whilst in internment during the First World War. There are three aspects to it: the precession of the equinoxes, the variable eccentricity of the orbit of the earth around the sun, and the axial tilt of the earth.

The precession of the equinoxes refers to the earth's slow movement backwards along its own orbit, so that for example on 21 March this year we would be looking at the sun from the very opposite end of the orbit to that on 21 March 13,000 years previously. This phenomenon also causes the stars we see at night slowly to rotate, as in the phrase "the dawning of the Age of Aquarius" – following the previous astronomical age, Pisces (the sequence runs backwards). The axis of the earth completes one full cycle of precession approximately every 26,000 years. At the same time the elliptical orbit rotates more slowly, on a cycle of 100,000 years. The orbit is close to a full circle, being only slightly elliptical, but the point of maximum distance from the sun in say the northern hemisphere summer does vary over time, due to the influence of the gravity of Jupiter and Saturn. The combined effect of the precession of the equinoxes and the variable eccentricity of the orbit is a 22,000-year cycle between the maximum and minimum insolation of either hemisphere. In addition, the angle of the earth's rotational axis oscillates between 22.1 and 24.5 degrees, on a 41,000-year cycle. It is currently 23.44 degrees and decreasing.

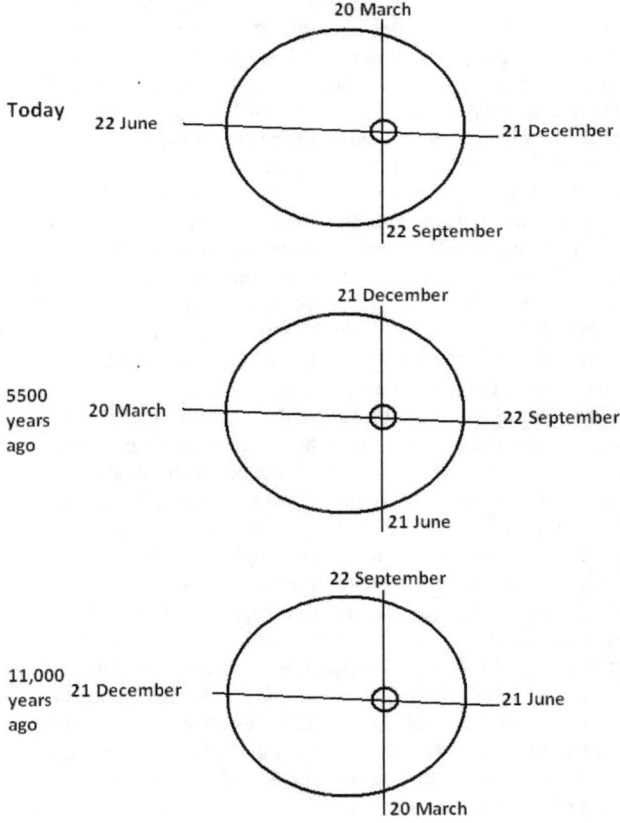

The combined effects of precession and the rotation of the Earth's orbit. Today winter in the Northern Hemisphere occurs when the earth is at its closest to the sun. 11,000 years ago it occurred when the Earth was at the far end of its orbit. The visible effect is that the first set of stars we see each night gradually changes.

The cycles have little or no effect on the total annual budget of heat from the sun (Tim Flannery quotes a figure of less than 0.1 percent, page 42). However they affect the distribution of heat over the planet and so the climate of the earth. For example in years when the three factors combine to give maximum insolation to the southern hemisphere, a proportion of that insolation will be reflected straight back out into space by the Antarctic ice cap, and so be lost. At the same time, the northern hemisphere would have a long run of the type of very cold summers needed to form an Arctic ice sheet.

There has been an increasingly good match between ice age temperatures dated from thorium isotopes taken from oceanic cores and the Milankovitch cycles, which had been abandoned as of little interest by the 1970s. There are good matches to the cycle in the peak glaciations at 100,000, 42,000 and 21,000 years ago. However, it is as well to remember that the cycles have no claim to cause the ice ages, but only to affect the cycles within the ice age. The Milankovitch cycles were there before the ice age ever began.

More recent thinking amongst the scientists is that it is much more likely that the ice age was brought about by a steady decline in world temperatures over 50 million years from the Paleocene, and that fall can probably be put down to the movements of the continents in the polar areas. Since the time of Milankovitch, Wegener's continental drift, or a form of it, has become accepted reality in geological circles (see next section), and it has an impact on the story of the Ice Ages.

As far as the onset of the Ice Age is concerned, the distribution of the continents at this time is critical, but there is an important difference between the northern and southern hemispheres. By seven million years ago, Antarctica had moved plumb over the south pole, and glaciation had begun in East Antarctica well before that – as early as 35 million years ago on some estimates. No ocean current could penetrate the area to warm it up. However the north pole is almost the negative image of the south. Instead of a continent surrounded by oceans, there is an ocean surrounded by continents. There is only one real entrance to this ocean, through the gap between Greenland and Norway. The back door – the Bering Strait between Siberia and Alaska, only 85 kilometres (58 miles) wide now – was sometimes completely closed when sea levels fell as a result of the formation of land ice. This would have prevented ANY cold Arctic water escaping into the Pacific, so that in turn warm

Atlantic water could not flow through the Arctic. If you leave the front and back door of your house open, a strong breeze will blow straight through. If you close the back door the breeze will drop to almost nothing. Warm water was blocked from the Arctic by cold water unable to move out of the back door. This effect still operates today. The Gulf Stream flows northwards to Norway, evaporating and becoming denser all the way until it sinks as it is heavy with salt – but if it could push the cold Arctic water out through the back door, it would carry on northwards, and there would be no Arctic ice sheet, though there would still be one on Greenland. Only a small fraction of the massive amount of heat transferred by the Gulf Stream would be sufficient to melt the Arctic ice cap.

In fact the complete closure of the Bering Strait is known to have taken place as several points in the ice ages, forming a land area known as Beringia. The current seas are so shallow that it is thought that this covered a very extensive area, 1000 kilometres (600 miles) from north to south. The emergence of such a complete barrier to oceanic currents would have had a feedback effect, reinforcing the ice age once it emerged from the seas. However this land bridge does not have to be 1000 kilometres (600 miles) wide to stop the currents – it need only be a yard wide! It is flooded now, and we are in an interglacial. It is known to have dried out in glacial periods. As it is in fact a shallow area of continental shelf, it could be flooded quite quickly, within a decade.

***********************************

The year 1953 saw the final resolution of a puzzle which had occupied many minds over the course of the previous two centuries – in fact going back to the very first geologist, James Hutton – the age of the Earth. This was achieved by taking a completely new approach. The problem is that due to the continuous recycling process which affects all rocks on the Earth, there is nothing on the Earth itself which dates back to the very beginning, although there are outcrops now known to be over three billion years old. The previous expert in this field, Arthur Holmes, had already come to the conclusion that the Earth was indeed three billion years old, but was having difficulty in persuading the rest of the world.

It was then that a young American man called Clair Patterson and his PhD supervisor hit on the idea of using meteorites, which they considered to be left-over materials from the very first days of the solar system. Not all meteorites fall into this category, but some of them do.

Patterson eventually obtained time on the very latest mass spectrograph at the Argonne National Laboratory in Illinois. When he first obtained his results, Patterson became so excited that he drove all the way home to Iowa and told his mother to send for a doctor as he thought he was having a heart attack. He was not, and was in due course able to announce an age for the Earth of 4.55 billion years, a figure not since challenged.

\*\*\*\*\*\*\*\*\*\*\*\*\*\*\*\*\*\*\*\*\*\*\*\*\*\*\*\*\*\*\*\*\*\*\*\*

The world of the geologists was to be turned upside down as a result of the development of paleomagnetism and the mapping of the ocean floor from the 1950s. Only then was the mechanism of plate tectonics revealed for the first time.

The mapping of the ocean floor had begun as early as 1872-6, when a Royal Navy ship, the *HMS Challenger*, set out to collect thousands of soundings. Although the overall pattern of the ocean floor did not emerge, the expedition found two significant things. The first of these was the existence of deep sea trenches. In fact the ship stumbled upon the deepest of them all, finding the Challenger Deep in the Marianas Trench at 11,138 metres (36,200 feet). The other discovery was the Mid-Atlantic Ridge. Sounding indicated a rise in the sea floor by as much as 4,600 metres (15,000 feet) above the abyssal (ocean-bottom) plain in the middle of the Atlantic.

The first hard scientific evidence for continental drift came in the 1950s from the study of the residual record of the earth's magnetic field stored in rocks, a subject known as paleomagnetism. (The prefix "palaeo" or "paleo" is common in geology and is taken from the Greek word for "old".) Certain minerals containing iron deposited in rocks lock in a record of the direction and intensity of the magnetic field when they form. Most notably, this phenomenon can be observed in basalt, the volcanic rock which underlies the ocean. When it cools past a certain critical temperature known as the Curie point, the iron mineral magnetite within it records the magnetic orientation of the earth frozen at that time. A sensitive device called a magnetometer was invented in 1956, which could measure this magnetism. Reading from rocks in widely separated regions only made sense – that is, pointed to the same north – if the continents were moved from their current positions to the places on the globe where they must have been when the rocks were laid down.

To the strange story coming from paleomagnetism came another strand of evidence, this time from the ocean floor. The invention the nuclear submarine and the opening stages of the Cold War made a much better knowledge of the contours of the ocean floor a requirement. Reliable mapping of this finally began to take place in the 1950s, when the development of sonar and seismic equipment made the physical plumbing of the depths – on occasions five miles down and more – as undertaken by the *Challenger* unnecessary. (Nevertheless, many cores of the sea bed itself have now also been recovered.) The evidence from the oceanographic surveys was collected and published in a map of the North Atlantic ocean bed produced by Bruce Heezen and Marie Tharp of Columbia University in 1957. This showed for the first time the true extent of the Mid-Atlantic Ridge, the most sensational geological discovery of the twentieth century and running the complete length of the Atlantic. It stands high above the abyssal plain, which itself is progressively older, more subsided and so deeper away from the ridge. In only one place does the ridge reach the surface, in Iceland.

Heezen and Clark realised the significance of the Mid-Atlantic Ridge immediately. For one thing, its profile closely resembled that of the East African Rift Valley, already thought to be a spreading crack in the surface of the earth. The notch at the top of the ridge was very similar, producing a pronounced rift valley – that is, a valley caused by faulting on either side of it – sometimes well over 1500 metres (one mile) deep, and 16-32 kilometres (10- 20 miles) wide .

The pair's next map of all the world's ocean floors, since widely reproduced, appeared in the *National Geographic* magazine in 1967. This map showed for the first time the existence of volcanic mid-ocean ridges running across all the world's oceans. In fact a map of these shows the world looking like a tennis ball. In any event, the new map fascinated the world. Drawn to an exaggerated vertical scale of twenty to one over the horizontal, it showed some totally unexpected features. These included massive fans of sediments carpeting the ocean floor off the coasts of India to depths of 19 kilometres (12 miles). Debris from the erosion of the Himalayas has been dumped into the Arabian Sea and the Bay of Bengal off the Indus and Ganges/Brahmaputra deltas, and has then spread out beyond the southern tip of India, on the eastern side by 3000 kilometres (2,000 miles). The undersea mechanism which can achieve this came to be known as a turbidity current. These are often triggered by earthquakes and can run at 72 kilometres per hour (45 mph), with a range far in excess of any land avalanche.

The Mid-Atlantic Ridge in Iceland

It was quickly understood that new ocean crust was being created along the mid-ocean ridges from within the mantle, the layer beneath the crust. Then by a process of sea-floor spreading, the new crust moved either way away from the ridge – in the case of the Atlantic, east and west. Hence South America, for example, is not drifting, but moving as part of a slab of crust or plate which is itself being newly created along its eastern border at the Mid-Atlantic Ridge. Now at last there was a theory that gave an explanation for the facts. This meant that every description of the processes of mountain building written before 1964 or so became immediately out of date.

Another part of the picture is the existence of mighty oceanic depths, the most famous of which is the Mariana Trench in the western Pacific, which is 11 kilometres (7.5 miles) deep – deeper than the highest mountain on earth (Everest is about 8 kilometres (5 miles) above sea level), but of the same order of magnitude, a fact which is not likely to be coincidental. In fact the total amplitude of the wrinkles on the surface of the earth is tiny, making it smoother than a billiard ball for its size, as they are constrained by the inward pull of the earth's gravity and the fluidity of its mantle and core. The highest known mountain in the solar system is Mons Olympus, on Mars, at nearly 22 km (14 miles). Although much smaller than earth, Mars has a solid interior which can support greater weights. Any mountain that size on earth would immediately sink into the mantle.

More evidence for plate tectonics arose from an unplanned source. During the late fifties and sixties, the Americans set up a worldwide network of seismographic stations, but the purpose of these was not to locate and measure earthquakes – it was to detect and measure the nuclear test explosions which were being conducted at that time by the Soviet Union, notably on the Arctic island of Novaya Zemlya. Nevertheless, these stations DID locate and measure whatever natural shaking of the earth's crust which was taking place at the same time. This revealed a pattern of earthquake activity worldwide, but the earthquakes were not taking place just anywhere. They were overwhelmingly concentrated in certain zones, some crossing the continents, others under the middle of the Pacific. The military scientists found that they had mapped the borders of the world's tectonic plates.

As the earth is not expanding, if oceanic crust is being created along the ridges, then it is also being destroyed elsewhere. In the western Pacific it is being dragged down underneath the east Asian plates, a process known as subduction. The deep oceanic trenches indicate its downward trajectory. It dives underneath Japan, causing the tremendous amount of earthquakes and volcanic activity there as the crust melts under the earth. Again, hard scientific evidence for subduction came from deep reflection profiles produced from seismic data, which showed the very image of oceanic plates dipping into the mantle on their way to destruction beneath the continents. Note that subduction is not a feature of every continental shore – for example both sides of the Atlantic are passive margins and lack submarine trenches. Subduction is only taking place in two small places, the Puerto Rico Trench and the South Sandwich Trench.

Within the Pacific basin, the East Pacific Rise is the eastern Pacific equivalent of the Mid-Atlantic Ridge, but it is a much smaller-scale affair, rising only about 300 m (one thousand feet) above the surrounding sea floor, with a barely detectable notch at the top. However, it is creating new sea floor much faster than the Mid-Atlantic Ridge. The sea floor here is spreading apart at a rate of up to 22.5 cm (nine inches) a year, compared to 2.5 cm (one inch) a year in the Atlantic. The East Pacific Rise is both hotter and more volcanic. The reason for the fast movement appears to be equally fast subduction all around the Pacific.

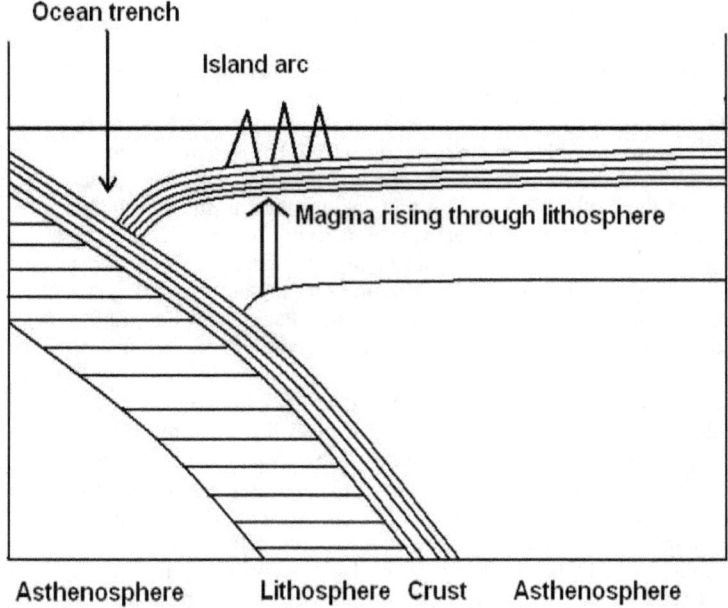

The process of subduction, where one oceanic tectonic plate dives beneath another, creating both an oceanic trench and an island arc. If a continental (rather than oceanic) tectonic plate (such as the Indian) crashes into another, the results are different – a chain of mountains, in that case the Himalayas.

\*\*\*\*\*\*\*\*\*\*\*\*\*\*\*\*\*\*\*\*\*\*\*\*\*\*\*\*\*\*\*\*\*\*\*\*\*\*\*\*\*\*\*\*\*\*\*\*\*

Paleomagnetism then made another contribution to the emerging picture of sea-floor spreading. The magnetic record showed that from time to time, the earth's magnetic field has reversed, so that south became north, and north, south. The last reversal was about 780,000 years ago. If the theory of sea-floor spreading was correct, then these magnetic anomalies should be recorded symmetrically either side of the mid-ocean ridges. Studies by two Cambridge geophysicists reported evidence of this very phenomenon on the Carlsberg Ridge in the Indian Ocean in the journal *Nature* in 1963, ever since remembered as the "Vine and Matthews" paper. Their observations were later confirmed from ocean cores as the big, well-funded American oceanic institutes got involved (deep sea drilling time does not come cheap!) The next step was to find dates for the magnetic reversals recorded in the rocks, to see if those found around the world coincided. A new mass spectrometer was constructed at the University of California which could measure tiny amounts of radiogenic argon, and this confirmed that the magnetic reversals of 990,000 years ago – the "Jaramillo Event" – were contemporary with each other. It also allowed the rate of spreading to be measured. The history of this scientific episode was recorded in a book by W. Glen called *The Road to Jaramillo*. Paleomagnetism as applied to ancient fragments of continental rock continues to extend the history of plate tectonics back in time.

Detailed study of the oceanic ridges showed something else – they are not straight as originally thought, but are strikingly displaced, as if two planks laid side by side had a line drawn across them and one of them had then been moved along. The American geologist Tuzo Wilson realised that the ridges were being moved along special faults, where no up or down movement took place. Slices of the ocean floor simply slide past each other. Wilson called these features transform faults.

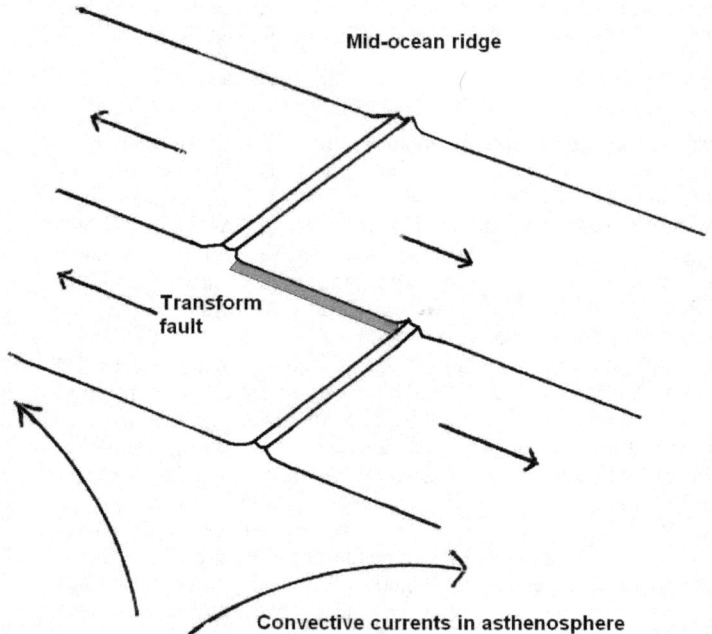

Structure of a transform fault

Fossil evidence has added to the picture of sea-floor spreading, and has the great advantage that it is much easier to examine a slide of a deep sea core than it is to book time on a mass spectrometer. The ocean floor is covered with shell debris, most of it the remains of single-celled floating organisms called foraminiferans and radiolarians. There is a difference between these, as the forams build calcareous shells, and live mostly in temperate and tropical waters. Radiolarians build siliceous shells and live in cold water, along with another type of plankton, diatoms. Both forams and radiolarians evolve quickly and can be used to date submarine deposits. An Eocene foraminiferan fifty million years old does not look like a modern one. The results from the cores once again confirmed the reality of sea-floor spreading – the further away from the mid-ocean ridge, the older the foraminifera which could be found in them. This evidence also showed that there is NO sea floor older than the Jurassic – everything over about 150 million old has been subducted – so the ocean floor is much newer than most of the continents.

The confirmation of plate tectonics proved a great boost to the methods of Charles Lyell – what is happening in the earth today must have happened in a similar way in ancient periods. Puzzling features such as the mechanics of mountain folding – which seemed to require at least one slab of crust to move – were resolved. Previously mysterious formations could be re-examined.

\*\*\*\*\*\*\*\*\*\*\*\*\*\*\*\*\*\*\*\*\*\*\*\*\*\*\*\*\*\*\*\*\*\*\*\*\*\*\*\*\*\*\*\*

The ocean floor has continued to yield new surprises as technology developed. Associated with the mid-ocean ridges themselves were later found new, strange phenomena – black smokers. These were only detected when illuminated by the lights of the submersible *Alvin*. When sea water permeates cracks in the open, molten crust under the ocean, it becomes superheated, to 300 degrees Celsius or more, but cannot boil because of the immense pressure. It becomes charged with minerals in solution, then issues from hydrothermal (hot water) vents, building fantastic chimneys of iron sulphide (pyrites), belching black smoke, and supporting an amazing and colourful fauna of tube worms, gigantic clams and other previously unknown creatures. The base of the food chain consists of bacteria which rely on chemical energy for their metabolism rather than sunlight and photosynthesis.

In fact the science of geology is still rapidly advancing. This is now a world of electron microscopes with secondary X-ray detectors and mass spectrometers. For example the study of "Paleoclimates" – the climates of ancient times – involves the very latest available scientific techniques, measuring minute quantities of the different isotopes of potassium, beryllium and other elements laid down in ancient rocks. One aspect of the new technology is the usage of zircon (zirconium silicate) crystals to provide accurate dates for rock sequences. Zircon is a very tough mineral formed in granite, which often survives erosion to reappear in tiny quantities in later sandstone sediments. Methods have been developed to extract these crystals, which contain a minute amount of radioactive lead. This provides a clock to date the formation of the original granite. It is due to the development of zircon dating that the ages of the geological periods are much better defined than they used to be. Forty years ago the base of the Cambrian was put at 600 million years, a figure now reduced to 542 million years.

# Chapter 15 – Twentieth-Century Astronomy

By the 1830s, the dimensions of the solar system were well-known, especially following two sets of observations already noted: Cassini's observations of the position of Mars viewed from Paris and French Guyana in 1671, and the measurements made during the transits of Venus in 1761 and 1769. The distance from the Earth to the Sun is in fact 150 million kilometres, defined as one "astronomical unit". The problem was that this implied that the "fixed" stars must be an almost unimaginable distance away, as they showed no relative movement in the night sky when measured from opposite ends of the Earth's orbit (say in December and again in June). This problem went all the way back to Copernicus, who concluded that the fixed starts were indeed unimaginably distant.

The apparent movement of an object in the night sky, when seen against a background of other stars, is called the parallax. The distance to a remote object can be calculated using simple trigonometry given a right-angled triangle, a base line of known length, and the appropriate angle of parallax, since the observer then has the length of one side and all three angles. Measured six months apart the baseline is 300 million kilometers, or 2 AU, which is certainly quite a long baseline to use in trigonometry.

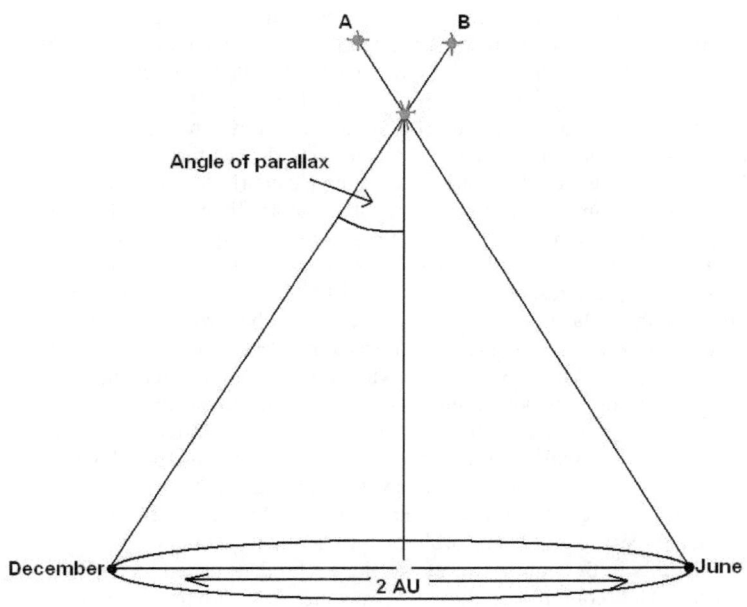

Parallax measurement. The baseline is two AUs long at opposite ends of the orbit of the Earth. In June the target star in line with distant star A, but in December it appears to have moved and is now in line with star B. The scale is exaggerated as no angle of parallax in the night sky is anything like this large.

However, by the 1830s instruments had been developed which were capable of measuring the tiny angles of parallax, which are measured in seconds of arc. A second, after all, is one-sixtieth of a minute, which is one-sixtieth of a degree, and there are 360 degrees in a full circle. In any event, a parallax second or parsec is defined as the distance to a star which shows an apparent movement of one second of arc from opposite ends of one-half of the Earth's orbit (one AU). If a star is one parsec away, then it will show an apparent movement of two seconds of arc from measurements taken six months (or 2AU) apart. It only takes a simple calculation of this to come up with a value of 3.26 light years for one parsec, and there is not a single star that close to the Earth. Copernicus had been right!

The pioneers in the new measurements selected stars with large proper movements, that is, movements across the sky observable over centuries (as opposed to no observable movement), which they thought must be relatively close. In 1838 a German called Friedrich Bessel came up with a parallax of 0.3136 seconds of arc for a bright star called 61 Cygni, giving a distance of 10.3 light years (now known to be 11.2 light years, so he got very close). In the same decade the Scot Thomas Henderson measured the distance to Alpha Centauri, third brightest star in the sky and in fact our nearest neighbour at 4.3 light years. The rate of parallax measurements from then on was roughly one a year until 1900. This speeded up when measurements made by eye using the crosswires of a telescope were replaced by measurements taken from photographic plates. Towards the end of the twentieth century, thousands became available from measurements taken by the satellite Hipparcos, which could swing on a wide orbit to obtain better angles.

Spectroscopy, as noted, allows information to be gleaned about the composition of the stars, using the spectral "bar codes" given off by the elements they contain, but there was no information available about the mass of a star. However a method was worked out to give mass by observing the light from binary stars, notably Alpha Centauri, which is in fact three separate stars orbiting one another. This is because of slight shifts in the light as the stars orbit each other, called the Doppler effect. This compresses light waves which are moving towards us towards the blue end of the spectrum, and has the reverse effect for objects which are moving away – the redshift. The size of the shift tells us the speed at which the object is moving, and this can be put into a formula which includes the rotational (angular) momentum of the individual stars to give the mass.

It was then a question of relating the brightness (later found to correspond to mass) of a star to its colour, and for different stars, these figures can be plotted on a graph which makes it clear that there is a definite pattern to the stars in the sky. The first person to produce such statistics was in fact a complete amateur, who was carrying out his observations as an unpaid observer, just hoping to gain experience. He did this at the University of Copenhagen, and his name was Enjar Hertzprung (1873-1967), by profession a photographic chemist. One of the things he found was that blue and white stars are always bright, whereas red and orange stars may be bright or faint – a very large red star (which is relatively cool) may still be bright. He published his results in a journal of photography in 1905 and 1907, where nobody from the scientific community noticed them. So it was left to an

American, Henry Russell (1877-1957) to make the same discovery – there is a relationship between the magnitude/brightness of a star and its colour – and to publish his results in the form of a graph (in 1913), now known as the famous Hertzprung-Russell diagram.

The colour of a star tells us its surface temperature, and this can be measured using just three wavelengths When the temperatures (or colours, since the two are closely related) are plotted against the brightness (or mass) on an HR diagram, they fall into a distinct band. Hot, massive stars lie at one end of this, our Sun is in the middle, and cool, dim and relatively small stars are at the far end of the main sequence. However there are two other groups, red giants (large, cool, bright) above the main sequence, and white dwarfs (dim, small, hot) below it. (A red giant is a star in the late stage of its evolution, when it has expanded massively whilst cooling. A white dwarf is a small, very dense star which has reached the end of its evolution.)

We have already met the next person on the trail of the stars, Arthur Eddington (1882-1944), just like Francis Dalton a Quaker originally from the Lake District (he was born in Kendal). During the First World War, although aged 34, he became eligible for conscription, and was certainly prepared to defy the authorities on the grounds of his Quaker beliefs, but the Government backed down and let him stay in his post at Cambridge, where he was a professor and director of observatories. After all it took a lot more money to train a scientist than it did a trooper! It was Eddington who undertook observations of the solar eclipse of 1919 to confirm Einstein's general theory of relativity.

In the 1920s Eddington gathered all the available data on stellar masses and combined it with the brightness data from the HR diagram, managing to show that the brightest stars on the main sequence are also the largest. The reason is quite straightforward – large stars have an immense inward gravitational pull, which forces them to burn fuel to counteract this tendency to collapse inwards, so the larger the star, the brighter it is. Through his calculations, Eddington concluded that the central temperature of any star on the main sequence, no matter what its size, is about the same – it must be controlled by the same process. By this stage it was fairly clear that the source of this energy derives from the conversion of hydrogen into helium within the star. Following Einstein's Special Theory of Relativity, which made it clear that very small amounts of mass convert into very large amounts of energy, Eddington came to the conclusion – not since doubted – that subatomic energy, which is abundant in all matter, could provide enough energy to keep the Sun burning for 15 billion years. He noted the research of

Francis Aston (co-inventor of the mass spectrometer and discoverer of the isotopes of neon) on the conversion of hydrogen into helium, the reaction which fires the Sun. The mass of the helium atom is slightly less than that of the four hydrogen atoms from which it is created, amounting to one part in 120 – but this represents an immense amount of energy, which is released into space from bodies like the Sun.

*****************************************

The next great steps in astronomy concerned the sheer size of the universe. A team from Harvard undertook the study of a class of pulsating stars known as Cepheid variables, which dim and brighten regularly, but with a period varying from perhaps only a day to a hundred days. Photos were taken in Peru and sent back to Harvard for analysis. A leading member of the team there, unusually for the times, was a woman, Henrietta Leavitt (1868-1921), who was also deaf. She saw a pattern in the pulsations of Cepheids in the Small Magellanic Cloud (SMC), a small galaxy associated with the Milky Way. This is a distant galaxy, beyond the Milky Way, and so far away that objects within it, though they may be separated by hundreds of light years or more, are all still more or less the same distance away as far as an observer on the Earth is concerned. Leavitt found that the brighter Cepheids within it pulsated more slowly, and in 1912 she constructed a mathematical formula for this property of the Cepheids, known as period-luminosity. According to the formula, and based on her observations, a Cepheid in the SMC with a period of three days is only one-sixth of the brightness of a Cepheid with a period of 30 days. To go any further, she needed to know the absolute magnitude of some nearby Cepheids, as astronomers had learned how to calculate distance based on absolute magnitude. It would then be possible to calculate the distance to the Cepheids in the SMC, and so the true distance to that galaxy.

In fact the data Leavitt required for calibration purposes was made available within one year, and by none other than Hertzprung, even if it wasn't very good data! The problem was the distorting effects known as extinction and reddening (not red shift), where interstellar dust interferes with the light from distant objects. A year later, as it were from the other half of the HR diagram, Russell made a correction for this. The first estimate was 30,000 light years to the SMC; the modern figure is 170,000 light years. Even at the first estimate, it was immediately obvious that the universe is a very big place. Such was the value of the technique, moreover, that the presence of a Cepheid in any remote cluster of stars could be used to obtain a fair estimate of the distance to the whole group. (One of the best-known of all stars is a Cepheid – Polaris, the Pole Star.)

The next breakthrough came with the construction of the first 100-inch telescope, named the Hooker after the man who funded it, on Mount Wilson near Pasadena in California in 1918. An observer called Harlow Shapley, at first using the 60-inch telescope at Mount Wilson, managed to establish some facts about the Milky Way, many of which had long been suspected. These were firstly that the Sun is not at its centre – that lies about 32,000 light years away; and secondly that it is about 325,000 light years across. Modern estimates for these figures are 28,700 and 90,000 and light years, so the second figure was quite a long way out, but it still meant that even our own Milky Way is large beyond human comprehension. (If it is wide, it is also remarkably thin, only about six hundred light years.)

Another famous telescope is located at the Lowell observatory in Flagstaff, Arizona. This had been set up and funded by one Percy Lowell, scion of a Boston business family. His early observations led him to claim that there are canals of the surface of Mars, crossing the entire planet, and so these Martians must be damn clever folk!

It was here, to great excitement, that a young man called Clyde Tombaugh found a brand new planet, Pluto, in 1930. This is now known to be located in the Kuiper Belt where there are in fact many similar if rather small objects (Pluto is smaller than the Moon). In 2006 Pluto finally lost its status as a planet.

The next man of note on the 100-inch telescope at Mount Wilson was Edwin Hubble (1889-1953), working initially in the period 1923-4. Some astronomers had begun to suspect that what had up to that point been described as "nebulae", clouds in space, were in fact whole galaxies, and Hubble was not only able to confirm this, but also to use Cepheids within them to calculate their distance. He came up with a figure for one of the closest, Andromeda, of 800,000 light years – ten times the typical distance to stars within the Milky Way. (This figure has since been adjusted to two million light years.)

Observations of the dying stars called supernovae have shown that one particular type, known as an Ia supernova, always explodes in the same way and at the same critical mass. These dying stars draw off material from a neighbouring star and when they reach a certain size, simply go bang, giving off as much light as their entire home galaxy over a period of about three weeks. These are called "standard candles" because they can be used as benchmarks against which to calibrate the brightness of (and so relative distance to) other stars and galaxies. By incorporating this data it became clear that not only are there billions of stars in the Milky Way, but there are billions of galaxies besides the Milky Way!

However Hubble is much better known as the man who discovered the red shift in the light from these distant galaxies. This is similar to the Doppler effect, whereby the light in objects moving away from the observer is shifted towards the red end of the spectrum, whereas light from objects moving towards the observer is shifted to the blue end. When observing the spectra of light coming from distant galaxies, it was found that the pattern of light and dark bands is identical to that coming from nearby stars, indicating that the same mix of elements must be present. However these bands are literally moved, in most cases towards the red, long-wave end of the spectrum.

Hubble was not the first person to notice this effect; that was a man called Vesto Slipher, working at the Lowell Observatory. By 1925 he had observed 39 red shifts and only two blue shifts. Assisted on the telescope by one Milton Humanson, a former mule driver, Hubble widened the search. He once again came up with a relationship: the further away the galaxy, the greater the red shift – in other words, the faster it is moving away. Reported in 1929, this became known as Hubble's Law. (No more blue shifts were found. One of them came from Andromeda, which is certainly heading our way.)

For Hubble this provided a relatively easy way to measure the distance to faraway galaxies – just measure the red shift in its light. Of greater significance, however, was the realization that this meant that the universe is expanding, just as Einstein's General Theory Of Relativity had predicted before he inserted the famous cosmological constant. In fact, Einstein did not give a fixed value to his constant, but offered a range of values which could slow down or speed up the rate of expansion (or, for that matter, of contraction). Several other mathematicians worked out their own versions of his equations on this basis. This also meant that the red shift is not strictly a Doppler effect, which is caused by an object moving away through space, but is caused by the space itself expanding, so that the light itself is stretched to longer wavelengths between the distant galaxy and the Earth. (Space is expanding like a balloon with dots painted on it to represent the galaxies. The galaxies are not travelling through space, so much as with space as it expands.)

The second implication of Hubble's law was that if the universe is indeed expanding, as it appeared to be, then it must have had a beginning, referred to disparagingly by astronomer Fred Hoyle (1915-2001) as the Big Bang. The first serious scientist to propose such a thing was the Belgian Georges-Henri Lemaître, who was also a priest, and who promoted the idea from 1931 onwards. However, few took these implications very seriously back in the 1930s and 1940s, because inaccuracies in Hubble's measurements (not fully cleared up until relatively recently) gave the universe an age of only about 1.2 billion years, younger than the then known age of the Earth itself, and less than one-tenth of the modern value of 13-16 billion years.

Instead there was a more general acceptance of Hoyle's Steady State universe, which envisaged an expanding universe with new matter – hydrogen atoms – being created to fill the gaps; rather than all the new matter being created at the very beginning. Under this model, the universe would always look roughly the same.

In 1948 a paper by Ralph Alpher (1921-2007) and Robert Herman predicted that the hot radiation emanating from the original Big Bang should still be detectable everywhere in the universe today, but that its wavelength would have stretched enormously from the original gamma rays and X-rays into long radio waves corresponding to black body radiation at a temperature of 5 K – five degrees above absolute zero. Whilst the controversy raged between the Big Bang, championed by George Gamow, and the Steady State, championed by Fred Hoyle, this paper lay neglected for 16 years.

However with the development of radio astronomy after the Second World War, it became clear that the ancient universe did not look the same as the modern one, as the Steady State team said it should. Signals were be picked up from far distant and previously unknown galaxies, called radio galaxies. Observations showed that these were much more abundant in the deep past of distant space – their density was greater, because the primeval universe was itself denser.

Another newly-discovered celestial phenomenon was the quasar, or quasi-stellar radio source, fist observed by radio telescopes in the late 1950s. These were found to emit massive amount of radiation from tiny areas of space. All of them featured extreme redshift, indicating that they must be very ancient, possibly forming not long after the Big Bang itself. At first there was great uncertainty about what they might be, but many were found. They are now thought to arise in the very active centres of young galaxies. They are completely different from anything which has been observed in nearby galaxies (the modern universe). The previously-unknown and distant radio galaxies and quasars meant some bad mornings over the bacon and eggs for Fred Hoyle – the universe has not always looked the same; far from it. (Indeed modern pictures of light billions of years old show lumpy galaxies which do not resemble the graceful spirals of nearby galaxies.)

In 1964 a Princeton professor called Jim Peebles independently came to the same conclusions as Alpher and Herman and had actually started to build a telescope sensitive enough to pick up the faint signals from the Big Bang. However, the physicist George Gamow had already suggested that the antenna at the Bell Laboratories at Holmdel, New Jersey – only 30 miles (50 km) from Princeton (and the birthplace of the transistor) – should be able to do the job.  Meanwhile, two radio astronomers were actually working with the Bell Labs antenna and found themselves constantly plagued by background radio noise, coming from all directions in the sky, and corresponding to black body radiation at three degrees above absolute zero!  These were Arno Penzias and Robert Wilson, and they did not know what they had found. News of this reached Peebles and his colleague Robert Dicke, who did know. In one of the most ridiculous awards ever, Penzias and Wilson were given a Nobel prize, which should have gone to Alpher and Herman. From this moment on, Big Bang was taken very seriously indeed.

Very precise instruments were built to measure the cosmic background microwave radiation and put into the COBE (Cosmic Background Explorer) satellite in 1989. The project was an American one, led by George Smoot and John Mather. In the first results announced in 1992, this measured perfect black-body background radiation at a temperature of 2.725 degrees K. COBE also mapped the property of the radiation known as anisotropy, minute differences in the direction of the radiation, measured at one part in 100,000. These are thought to indicate density differences or ripples in the early universe which "crystallised" into the later galaxies and areas of space without galaxies.

***********************************

Even before the Second World War, considerable progress had been made in developing an understanding of the chemistry of the universe. It was not realised until the 1930s that a body like the Sun is composed mainly of hydrogen, which is converting into helium with very few other elements present, and none in great quantity. The German physicist Hans Bethe was the first to describe the process of nuclear fusion which operates inside the Sun, where the heat and energy is available to fuse four hydrogen nuclei into one of helium. However a second process was identified, thought to be dominant in stars 1.3 times larger than the Sun. This is known as the (stellar) carbon or CNO cycle and requires the initial presence of a quantity of carbon, which acts as a catalyst in the reaction. As protons are added to carbon-12 it becomes a series of unstable nuclei – nitrogen-13, carbon-14, oxygen-15 and nitrogen-15. With the addition of this fourth proton the nucleus fires off a whole alpha particle and reverts to its former status as carbon-12! As an alpha particle is a nucleus of helium-4, the cycle has managed to convert four protons into one helium nucleus, along the way emitting large amounts of energy and two positrons (positive electrons) which are immediately annihilated by electrons. This process amounts to the transformation of what is in effect a single nucleus in a continuous loop.

Also during the 1930s came a contribution from George Gamow, (1904-68), a Russian who had defected to the west in 1933 and who worked at the George Washington University in Washington DC. He proposed a process known as "tunnelling" by which alpha particles (helium nuclei) could escape from the nuclei of radioactive atoms. They should be constrained within the nucleus by the strong nuclear force, but as they are observed to escape, they must have got out somehow! Gamow drew on quantum uncertainty as the source of the extra push of energy these particles need to get away. Each particle would borrow energy from quantum uncertainty, in that the amount of energy available is uncertain! (It's the crazy world of quantum mechanics again.) It would then repay the borrowing immediately on its escape, so no one would ever know. These ideas went into the thinking about the evolution of stars and their energy.

In the postwar years Gamow put his bet on the Big Bang, and set his PhD student Ralph Alpher (1921-2007) to discover how the heavier elements could have been created from whatever was there when the whole universe could be fitted into something the size of the solar system. The first atom bombs had by this time been dropped, and nuclear reactors were being built, and it had become clear that it is possible to make one element out of another, given enough energy. This had been observed to happen by neutron decay, where the mass and energy of the neutron converts to an extra proton and an electron, giving off a blast of gamma rays as it does so. Assuming that there was initially nothing but neutrons, Gamow and Alpher surmised that these would form the simplest element, hydrogen, and then the next, helium, with various intermediate stages including heavy hydrogen (which has one neutron, as opposed to no neutrons). They concluded that the common elements do indeed form in this way.

When it came time to submit this work as a scientific paper in the *Physical Review* of 1 April 1948, Gamow whimsically decided – rather to the chagrin of Alpher – to add another name to it, that of his friend Hans Bethe, who had had nothing to do with it. Thus the authorship reads Alpher, Bethe, Gamow (alpha beta, gamma).

Alpher and Robert Herman continued to work on the physics of the Big Bang. They soon concluded that there is a problem in getting the whole array of chemical elements out of the conditions thought to exist at that time, because although it is easy enough to generate vast amounts of hydrogen and helium, plus a small amount of lithium, the next stage is more difficult. The next element in the chain is beryllium-8, but in the conditions of the Big Bang, this immediately decays back to two nuclei of helium-4. In other words the elements heavier than lithium could not have been created at the Big Bang, but as they clearly originated somewhere, then where? Arthur Eddington famously suggested that if critics should say that the stars were not hot enough to effect transmutation, then "we tell them to go and find a *hotter place*" – presumably, to go to hell.

A contribution to this debate then came from Fred Hoyle in 1953. He suggested that carbon-12 could be formed from three helium nuclei (omitting the unstable beryllium-8 stage) if it possessed a property which he called "resonance". By this he meant an enhanced state of energy, and he suggested that experiments could be devised to test for resonance. Such experiments were conducted successfully by the American Willy Fowler, showing that the three alpha particles could merge together instead of (as previously predicted) smashing apart; this is called the triple alpha process. In fact, as early as 1917, it had been observed that elements with even atomic numbers – 2, 4, 6, 8 etc – are far more abundant than those with odd atomic numbers, which leant weight to the idea of their creation by some process of nuclear fusion.

This was a key step in understanding the manufacture of the heavier elements, with further additions now possible from carbon up to iron. Here the scientists found another road block, because iron is a very stable form of matter, held together with a minimum of energy. Huge amounts of energy would be required to produce heavier elements such as gold and uranium. The conclusion was reached that this would only happen at the very end of the life cycle of a star, when it undergoes gravitational collapse into a dense ball and then explodes as a supernova. This is a rare astronomical event which creates as much light from one star as otherwise comes from a whole galaxy. This would both make the heavy elements and scatter them in gas clouds through space. This all may seem like a wild idea, but observations confirmed it, notably during the supernova explosion in the Large Magellanic Cloud in 1987 which was observed in great detail. Another famous if not quite so widely observed supernova created the Crab Nebula in the year 1054.

Supernovae are now regularly detected in distant galaxies, at the rate of two or three a year, and have an importance beyond the merely spectacular as they act as standard candles against which to calibrate the relative distance to other stars and galaxies.

All that is left at the centre of a supernova is a super-dense core of neutrons – a neutron star. Only one year after James Chadwick discovered the neutron, in 1934, Fritz Zwicky (a Bulgarian) and Walter Baade suggested that a neutron star would be the outcome of a supernova. The neutrons would form because the electrons and protons would fuse together into electrically neutral particles. One estimate is that the weight of one teaspoonful of a neutron star would be 90 billion kilograms. In 1965 a neutron star was observed in the Crab Nebula, left behind after the great explosion.

Increasingly powerful telescopes and later space probes have led to the discovery of other new phenomena in space. One of these is the pulsar, a form of neutron star, the first of which was observed in 1967. This is a star which emits a powerful radio beam which appears to pulsate. As the star is rotating, the signal is only received when the beam is pointing towards the Earth, very much in the manner of a lighthouse.

Meanwhile a greater understanding of black holes has developed. A black hole itself is defined as an area where the force of gravitation is so strong that not even light can escape. Normal stars can and do orbit black holes, but if an object should fall into the hole, then it passes through the "event horizon" beyond which no escape is possible – the point of no return. At the centre of the black hole lies a singularity when all normal equations of physics break down. A great deal of the theory of black holes has been expounded by the English physicist Steven Hawking (1942-), whose book *The Brief History of Time* (1988) must have been the least well understood best-seller of all time.

Recent research at the Lawrence Berkeley National Laboratory by a team under Saul Perlmutter came to the startling conclusion that the expansion of the universe is not slowing down – the expected result –

but speeding up! Their conclusion is that some form of anti-gravity, an unknown force which is stronger than gravity and which is given the name dark energy, must be responsible. Perlmutter received shares in two Nobel prizes for his work, in 2006 and 2011. Other recent astronomy has attempted to measure the curvature of the far limits of space. If curvature can be found, then this will be further evidence that the universe does have a boundary, in fact the current edge of the explosion following the big bang. However, no sign of curvature could be detected – the researchers found a flat universe, meaning an infinite one. It appears to go on forever.

Note that whilst the expansion of the universe and the Big Bang are accepted facts, it is impossible to say what happened before the Big Bang, which might be defined as the ultimate singularity, since not even time existed at that point. However it is to be noted that Einstein's equations admit of either an expanding or collapsing universe, and some theoreticians take the view – which it is rather difficult to prove or disprove – that the current expansionary phase of the universe will eventually be followed by a gravitational collapse, and we shall end up with another Big Bang when everything will start all over again.

*********************************************

Back on planet earth, twentieth-century chemists found something exciting to do – to create brand new, synthetic elements. The strong nuclear force is only strong enough to hold together a maximum of 92 protons, the number found in uranium, which is in any case unstable. However new and of course even more unstable elements can be created by bombarding uranium with radiation of various types. The first element created in this way was neptunium, atomic number 93. By 1941 the next element had been created – plutonium, number 94, at the University of California in Berkeley. The man in charge of the research was the American Glenn Seaborg (1912-1999), who is also credited with the manufacture of another four new elements, and who had a hand in the development of several others.

It was predicted that plutonium would have fissile properties which would make it much more suitable for an atom bomb than uranium itself, only a few pounds of it being required to set off a chain reaction. Hence plutonium was then developed as part of the Manhattan Project which constructed the first atomic bombs. The first atomic bomb, Little Boy, which fell on Hiroshima, had a core of uranium 235. However the second one, Fat Man, which fell on Nagasaki in August, 1945, had a much smaller plutonium core. Rarely can a scientific breakthrough have had such a sudden, dramatic and terrible effect.

# Chapter 16 – Computer Science

The history of modern computing starts in an unexpected place – weaving. In 1801 the Frenchman Joseph-Marie Jacquard developed a loom which used punched cards to store its weaving instructions. Hence it could be "programmed" to produce different designs for different sets of punched cards. This same type of card – or a variant of it, punched paper tape – was then used by virtually every form of computer or calculating machine right up until the 1960s as the basic means of entering program instructions and data. Note that this is a very important distinction. Programs are sets of instructions for manipulating data, and they come in two overall types – operating systems, which organize the physical equipment of the computer, and applications, which do the useful things as required by the computer user. Applications depend on input from the user – his data. Data and programs are fundamental to computers but they are completely different things. Programs are referred to as software. The physical equipment – disc drives, memory, printers and so on – is referred to as the hardware.

Computing machines really begin with the English mathematician and inventor Charles Babbage (1791-1871) who produced a calculator called a Difference Engine, and then a design for a more sophisticated computer which he called an Analytical Engine (1837). This required thousands of moving parts which could not at that time be manufactured within the tolerances he required. A number of Difference Engines (mark II) were built from 1855 onwards, and a brand new one was built in 1971 to Babbage's original design, for the London Science Museum. Babbage was assisted in his work by Countess Ada Lovelace, daughter of Lord Byron. Her notes on his machine include what is considered to be the world's first algorithm meant to be processed by machine, so she in turn is seen as the first computer programmer. The modern computer language, Ada, is named after her. An algorithm is a set of instructions meant to be obeyed sequentially which will process data. (A single computer program may be very large and may contain different algorithms within it.)

In the later 1880s the American Herman Hollerith (1860-1929) then came along with the idea of using punched cards for data as well as machine instructions. He invented a key punch to make the holes in the cards and a tabulator, a machine to read the cards and summarise the data on them. This was the foundation of modern data processing. The machines were used to process the vast amount of data generated by the 1890 United States census, a task which was consequently completed years ahead of the previous schedule.

Calculating machines of increasing sophistication were then developed, notably by the American company International Business Machines (IBM), which had evolved from Hollerith's original business. However what today we would call a computer, which stores instructions in its own memory, had to await the urgent imperatives of the era just before and into the Second World War. The most impressive machine of this era was the Colossus, produced under the British Post Office engineer Tony Flowers, and little known for many years because it was an official secret. It was used to crack the German military codes at Bletchley Park in Buckinghamshire. One of the most influential thinkers in the contemporary computer business, Alan Turing (1912-1954), also worked here and helped to design another computer, the bombe. Turing achieved a definition of the computer known as "Turing-complete": a machine that can process a set of instructions sequentially, but branching on conditions as required (IF...THEN....GOTO). Turing famously committed suicide at the age of only 41, unable to deal with his homosexuality.

By 1945 the Hungarian-born American mathematician John von Neumann (1903-57) had achieved the definition of modern computer architecture. The computer should be an electronic, digital machine using the binary code. It will contain a central processing unit, which will contain an arithmetic and logic unit (ALU) to carry out arithmetical and logical operations, and processor registers to hold the data currently being processed. This will also contain a control unit to fetch the next instruction, instruction registers to hold those instructions, and a program counter to step through the program. The computer memory will sit separately from the central processing unit and will store both data and instructions. The computer will communicate with mass storage devices by means of input and output devices. Data will be stored on mass storage devices (such as stacks of punched cards, later magnetic media) and will be passed into the computer memory before finally being passed to the process register. In short the Von Neumann architecture describes a stored-program computer.

The first of the modern general-purpose electronic computers was the American ENIAC (Electronic Numerical Integrator and Computer). This machine featured high-speed electronics and was fully programmable. However, it did not have a program stored in its memory, and as it used the decimal rather than the binary system, it did not represent the future. (The programming instructions had to be set up using switches and patch cables before the start of each run, and a team of six women was employed for this purpose.) The ENIAC could perform 5000 arithmetic instructions per second. It came into operation at the end of 1945. However, it was no laptop. It weighed 30 tons and contained over 18,000 vacuum tubes and 1500 electric relays, and consumed 200 kilowatts of electric power when it ran. However it was a very successful machine, being in almost constant use for the next ten years at the University of Pennsylvania.

The first stored-program computer, the Manchester "baby" or Small-Scale Experimental Machine, was up and running at Manchester University in England in 1948, but this was very limited in scope. Nevertheless it was used as the platform for the first commercial computer, the Ferranti Mark I, which was installed at the same university in 1951. The following year, IBM announced its first commercial computer, the 700 series. The company had completely underestimated the new machines at first, predicting a worldwide market of only six computers, but it rapidly became the leading manufacturer of mainframe computers.

Just as the Second World War had provided the impetus to develop computers, the Cold War between the United States and Russia also stimulated the next really large projects and accompanying developments in hardware. A system called Project Whirlwind was originally developed from 1945 in the USA to enable the same flight simulator (used for training new pilots) to be used for different models of aircraft. However within a few years this had morphed into a massive system to monitor the defence of the whole of the United States against air attack, known as the SAGE system. This was implemented in the early 1950s at 23 centres, at each of which was installed two IBM AN/FSQ-7 computers. Each weighing 250 tons, this was the largest computer ever put into service.

By the end of the 1950s, two standard high-level computer languages had appeared, which enabled programmers to write in a form of English instead of in machine code instructions. The first of these, Fortran, used for scientific purposes, came originally from IBM, but rapidly spread to other manufacturers. Amazingly, it is still in use today. The other, Cobol, was a language used for commercial and administrative purposes. Sponsored by the US government itself, it reached a peak in the 1970s before fading away. Incidentally, this was the first language I learnt in my own computer career, having also studied some Fortran as a postgraduate student. Another new language, BASIC, was developed at Dartmouth College in the USA and put into use by 1964. Designed with simplicity in mind, it quickly caught on and was rapidly adapted to the first microprocessors.

From 1955, core memory was introduced, greatly speeding up the random access memory of computers. This dominated hardware for about 20 years, until it was replaced by semiconductors. Also in 1955, transistors were introduced to replace vacuum tubes, greatly reducing the size and electricity consumption of monsters such as ENIAC. Soon to follow was the era of semiconductors (silicon chips). At first there were relatively few components on each chip, but according to one of the co-founders of Intel, Gordon Moore, that would double every year. In fact that turned out to be about once every 18 months, but nevertheless, by the twenty-first century, an integrated circuit could contain millions of components.

The first commercially successful minicomputer, Digital Equipment Corporation's PDP-8, appeared in 1964. This was soon to be found in every university computing department. By 1975, the first microprocessor-based computer had appeared. This was the Altair 8800, developed, of all places, in Albuquerque, New Mexico, by a hardware salesman called Ed Roberts. He was joined in New Mexico by Paul Allen, who had graduated, and by his friend Bill Gates, who left Harvard – much to his parent's disgust – to be one of thousands of young men creating nascent microcomputer businesses. Only a few of those businesses would become successful. Paul Allen and Bill Gates soon left to set up their own company in Seattle, Microsoft; Gates was easily to confound his middle-class parents by becoming a dollar billionaire at the age of 31.

The first successful personal computer, the Apple II, produced by Steve Jobs and Steve Wozniac, saw the light of day in 1977. In the meantime, useful applications were being developed to make the personal computer useful not only in the home, but as a business machine. The first of these was VisiCalc, a spreadsheet, which became an immediate success when it was launched at the end of 1979. The first modern-looking word processor, WordStar, appeared in the same year.

IBM had again missed the potential of this sector, but with the aid of two then-small companies, Microsoft for the operating system and Intel for the chips, it soon came to dominate this new market, introducing the IBM Personal Computer in 1981. (The President of IBM, John Opel, actually knew the mother of Bill Gates of Microsoft, as they both sat on the same company board at United Way.) Thereafter, almost all personal computers apart from Apples were made to the same model – "IBM compatible". Apple had enough originality, style and reputation to remain independent, producing its next successful machine, the Macintosh, in 1984.

*******************************************

The above gives a brief history of computers, but how do they actually work? Those readers who suspect that they are about to be baffled by the next section are allowed to move directly to the next chapter!

Computers use a binary code for both their data and their instructions. Binary numbers are loaded into the "registers" in the central processing unit of the computer, one "word" (32 binary digits or "bits" in modern machines) at a time to be processed. So how do binary numbers work? There are only two possible values, 0 and 1, yet any number can be represented in binary. Obviously this requires a lot more digits for a large number than a decimal system does, but then, the basic unit is 32 bits long, and that number of bits can still contain a very large number.

Binary numbers represent increasing powers of 2 to the left, as decimal numbers represent increasing powers of 10 to the left.

Binary 1 = decimal 1
Binary 10 = decimal 2
Binary 11 – decimal 3
Binary 100 = decimal 4
Binary 101 = decimal 5
Binary 110 = decimal 6
Binary 111 = decimal 7
Binary 1000 = decimal 8

In terms of binary arithmetic:

1+0 = 1
1+1 = 0 carry 1

So to add 3 + 5:

0011 +
0101=
1000

In order to manipulate binary digits, the computer recognizes them simply as a state of "switched on" (1) or "switched off" (0). It then applies Boolean logic to these two states. Boolean logic was invented in 1847 by the English mathematician George Boole, who recognised that in this simplified system there are only two values, TRUE (on, 1) and FALSE (off, 0). The American Claude Shannon noted in the 1930s that there is an exact correspondence between Boolean values and electronic circuits, or logic gates, and these lie at the heart of any computer.

There are three primary Boolean operators, logical AND, OR and NOT, and their values for any combination of two inputs A and B can be displayed in a "Truth Table". In the case of the logical AND, both inputs must be true for the output to be true.

LOGICAL AND:
A....B.....Result
1....1......1
1....0......0
0....1......0
0....0......0

In the case of the logical OR, if either input is true, then the output is true.

LOGICAL OR:
A....B.....Result
1....1......1
1....0......1
0....1......1
0....0......0

Logical NOT requires only one input, which is reversed in value:

LOGICAL NOT:

A…..Result
1……..0
0……..1

In addition to these basic operators, there are others which might be described as derivatives. The most important of these is the logical NAND, which is NOT and AND combined, and gives the opposite result to AND: TRUE only if both inputs are negative.

LOGICAL NAND:
A….B…..Result
1….1……0
1….0……0
0….1……0
0….0……1

Then there is the EXCLUSIVE OR, which gives a value of TRUE if the inputs are different, and FALSE if they are the same (in other words either but not both inputs must be true):

EXCLUSIVE OR:
A….B…..Result
1….1……0
1….0……1
0….1……1
0….0……0

These basic operators can be used to carry out arithmetical calculations in the "arithmetic and logic unit" which sits at the heart of any computer. For example, the straight addition of two bits is represented by EXCLUSIVE OR. Adding 1 + 1 gives a value of 0 (carry 1).

One feature of both NAND and EXCLUSIVE OR, normally put to use in the case of NAND, is that it is perfectly possible to construct any other logical operator exclusively from NAND operators. This makes the construction of circuits simpler, because there is no need to use any other kind of logic gate. To take the simplest first, NOT can be achieved by splitting the input into two, and putting it through a NAND gate:

LOGICAL NOT

A....A.....Result
1....1......0
0....0......1

With LOGICAL OR, the two inputs go through two NAND gates in succession, using the output from the first NAND twice:

First NAND
A....B.....Result
1....1......0
1....0......0
0....1......0
0....0......1

Second NAND
0....0......1
0....0......1
0....0......1
1....1......0

For logical AND, first negate each input by splitting it and feeding it into a NAND gate, as in logical NOT derived by NAND above:

A....A.....Result
1....1......0
0....0......1

B....B.....Result
1....1......0
0....0......1

In this way the original truth table becomes:

A....B..........New A....New B
1....1.............0...........0
1....0.............0...........1
0....1.............1...........0
0....0.............1...........1

Then put this new input through another NAND gate:

| New A | New B | NAND |
|---|---|---|
| 0 | 0 | 1 |
| 0 | 1 | 0 |
| 1 | 0 | 0 |
| 1 | 1 | 0 |

Computers recognise two different forms of number, integer and floating point. Integers are simply the binary equivalents of the decimal number, and are held in 31 bits, with the final bit used to represent the sign. Floating point numbers are held in three sections: from left to right the sign bit, 8 bits for the exponent and 23 bits for the mantissa. For example the number 8 would be held with a sign of 1, an exponent of 11 (binary 3) and a mantissa of 10 (binary 2): 2 to the power 3. (However note that numbers are normalized before being put into floating point format, that is, all digits are moved to the right of the decimal point and the exponent incremented accordingly). Note that floating point arithmetic, which includes all fractions (i.e. non-integers) is not completely accurate. This is also a problem in the decimal system which cannot reflect some values such as one-third, Pi or the square root of two with complete accuracy. However for most calculations, floating point is accurate enough, especially when dealing with very large numbers such a light years.

Alphabetic characters, numbers, punctuation marks and control characters are also held in binary codes. The most important of these is the ASCII code (American Standard Code for Information Interchange), in which each character is allocated 8 bits (four characters to a 32-bit word). For example the letter A is represented by decimal 41, the letter B 42, and so on. 8 bits gives a possible maximum of 256 different symbols. Another code is EBCDIC (Extended Binary Coded Decimal Interchange Code), used on IBM mainframes and compatible machines. In fact a great deal of computing is properly data processing, and much of the data is in alphabetic format – names and addresses, for example. These codes are fundamental to data processing.

# Chapter 17 – Atmospheric Chemistry

First, what is a greenhouse gas? It is one which admits the rays of the sun, which arrive on earth in short waves, but does not let them out again when they are radiated back out into space as long waves. The most contentious and in some ways surprising of these gases is carbon dioxide. As long ago as 1896, the Swedish chemist Svante Arrhenius, the man who is credited with the discovery of ionic bonds, suggested that the burning of fossil fuels had the potential to raise temperatures globally. In fact, carbon dioxide only absorbs outgoing radiation at wavelengths of above 12 microns and a small amount of gas present in the atmosphere – such as was present before the Industrial Revolution ever began – captures all the radiation available at those wavelengths. Increasing the concentration in experiments has NO direct effect (see Tim Flannery, page 27). Moreover, the amount of the gas present in the atmosphere has increased from 280 ppm (parts per million) in the pre-industrial age to 380 ppm today, yet the climate was as warm in the Medieval Warm Period (approximately 950-1250 AD) as it was in the period 1960-90. An increase in concentration of 35% ought to have made a difference! Even today, with the amount of the gas increasing every year, the evidence for global warming is weak after 1997.

Yet the new science of geochemistry tells us that whenever the earth has become a hot place in the geological past, this has been reflected by high concentrations of carbon dioxide in the atmosphere – but is this a cause, or an effect? There is strong evidence that low temperatures produce low levels of carbon dioxide, for one simple reason – cold seas are biologically much more productive than warm ones. At the base of the food chain in the cold waters are phytoplankton, which can grow into great blooms if sufficient nutrients are present, and these draw down massive amounts – billions of tonnes – of carbon dioxide from the atmosphere. So when cold water spreads, so do the phytoplankton, and atmospheric carbon dioxide falls. When warm water spreads, the opposite happens and carbon dioxide is released back into the atmosphere.

In recent years, another form of feedback has emerged from cores extracted from the Antarctic ice. These and other sources have shown that the carbon dioxide content of the atmosphere was very much lower at the peak of the last glaciation, 18,000 years ago – only 200 parts per million, as opposed to 280 ppm in the pre-industrial age, and 380 ppm today. Of course this may well have reduced the warming, greenhouse effect of carbon dioxide, but was it a cause of the ice age, or an effect? There is strong evidence that is was an effect. The ice ages were very dry, and dusty, iron-rich winds blew off Patagonia across the Southern Ocean, greatly enhancing the supply of iron which phytoplankton need to grow. The result was massive blooms, reflected in the fossil record, which would have drawn down an estimated two billion tonnes of carbon dioxide from ocean, which would have then replenished itself with dissolved carbon dioxide from the atmosphere. So once an ice age got going, it would be greatly intensified by this effect, even if it is doubtful that it alone could have accounted for all the drop in carbon dioxide levels. Nevertheless the levels of carbon dioxide in the atmosphere did rise and fall in lockstep with the global temperature.

The same kind of result has come from measuring the level of dimethyl sulphide in the ice cores. This chemical is produced by marine algae and is thought to play a major role in the seeding of clouds. The measurements show that the output of DMS was nearly five time greater in an ice age than otherwise. The algae must have been doing well, and moreover, there must have been a lot of cloud! – reflecting away the sun. DMS is the chemical which gives us the smell of the sea.

The composition of the air is approximately 78% nitrogen, 20.9% oxygen, 0.9% argon, 0.038% carbon dioxide and then tiny percentages of other gases including water vapour, methane, ammonia, nitrous oxide and sulphur gases. This improbable mixture displays what scientist call a reduction in entropy, or a persistent disequilibrium, unlike anything else in the solar system. The Second Law of Thermodynamics predicts that without external inputs, systems become disorganized and are then said to possess a high degree of entropy, or state of disorganization. It is apparent that the atmosphere of the earth is highly organized. This is because this mixture is entirely the result of the organic processes taking place on earth. If there was no life, it is likely that the atmosphere would be composed of carbon dioxide, as it is on Mars. Furthermore the proportion of gases is not at all reflected in their importance to the cycle of life. Nitrogen and argon act as more or less inert fillers, bulking out the oxygen, which would become dangerous – it is highly combustible – at higher levels.

One of the most original thinkers in this area is the chemist James Lovelock (1919-), whose first contribution was a book called *Gaia – A New Look at Life on Earth*, first published in 1979. Lovelock maintains that the mass of biota upon the earth act to keep the earth's atmosphere in or close to its current equilibrium, so that, for example, if the amount of oxygen rises, then reducing gases (which inject hydrogen into the atmosphere) such as methane will be released, over time, to bring it down again (see below). Gaia is a system of controls and feedback mechanisms which always work to ensure that the planet maintains a comfortable environment, and especially temperature, for life.

Methane is twenty-four times as potent as carbon dioxide as a greenhouse gas. The biota of the earth produce a great deal of methane – much of it is generated by anaerobic bacteria (living with no access to oxygen), but a certain amount of it famously coming from the farting of cows. It also arises from the rotting of vegetable matter. Photosynthesis breaks down carbon dioxide ($CO_2$) into carbon, which is used by the plant, and oxygen, which is released. In chemical terms, the carbon dioxide is reduced to carbohydrate by the internal mechanisms of plant cells. Some of the carbon is turned into glucose to provide the plant with energy, and some of it is built into the structure of the plant, for example to make a tree trunk. When the tree dies, this carbon can be buried or "sequestered" – it is a carbon sink. This carbon can be released over the short term by the tree rotting in aerobic (oxygenated) conditions, or held over the very long term to form coal. At some point it is likely to reappear in the atmospheric cycle as methane ($CH_4$). However, methane is not stable in the atmosphere of the earth. It reacts with free oxygen to form carbon dioxide and water vapour.

$$CH_4 + 2O_2 = 2H_2O + CO_2$$

The only way that the release of buried methane could change the climate would be by it acting indirectly to produce large volumes of carbon dioxide and water vapour. In fact the role of methane in the atmospheric system is thought to be to keep the level of oxygen down by taking it out of the system in this way. Otherwise the free oxygen released by plants in photosynthesis would build up to dangerous levels. So in fact the atmosphere NEEDS methane, which is simply returning the carbon which the plants took from it in the first place.

The earth's processes also produce a great deal of ammonia ($NH_3$), in fact about 1,000 million tonnes of it a year. It is a waste product for many organisms, secreted in the form of urea or uric acid. Again it is a tiny constituent of the atmosphere only because it is unstable as it too reacts with oxygen.

$$4NH_3 + 5O_2 = 4NO + 6H_2O$$

As it releases so much reducing gas (hydrogen) into the air, its principal function is thought to be to reduced the acidity of the rain to a pH level of about 8, the optimal level for life. Without ammonia this pH level would drop close to 3, about the same acidity as vinegar. Imagine what would happen if the rain came down as vinegar. Another way to look at this is to ask, if the atmosphere is 78% nitrogen, where did it all come from? It is not like that on Mars. The answer must be that it is what is left over after the (organic) ammonia has reacted with oxygen to form water vapour.

At any event, as Lovelock points out, any outside observer analyzing the chemistry of the earth's atmosphere from a distance – as we do for Mars and Venus – would immediately conclude that the earth must contain life. Methane and ammonia must be being produced in truly vast quantities for there to be any present at all in the air, as both are so unstable in the presence of oxygen. In fact when asked what he would do to look for life on Mars, in the form of experiments to put on a landing probe, Lovelock scratched his head and replied that he would look for a decrease in entropy, because this would imply that something was out there causing such a state. The fact that the atmosphere of Mars IS composed of carbon dioxide was for Lovelock evidence enough that there was no need to look for life, because it meant that no decrease in entropy could be observed, before any landing was made on the planet. The gases which come from living things – especially oxygen, but also methane and ammonia – were not detectable.

Back in 1979 biologists assumed that life had adapted to the atmosphere of the Earth rather than created it, and Lovelock's ideas met strong opposition. (He himself notes the trajectory of new ideas: first dismissed as absurd, then maybe true, and finally we knew it all along.) Nowadays no one doubts that the atmosphere is a product of life, but to take it a further stage, Lovelock believes that it has evolved with life over billions of years, to protect life. The sun is thought to produce 30% more energy now than it did when the earth was young. At that time the atmosphere would indeed have been mainly carbon dioxide, which would have kept the earth warm. Over time the atmosphere has changed in composition, all the while maintaining similar temperatures. The idea is that the earth has managed to keep cool as the sun has become hotter by the removal of atmospheric greenhouse gases.

However, if you bought such a thermostat as Gaia, you would soon want your money back. It would be no use if it resulted in the disappearance of 90 per cent of the marine life it was supposed to keep comfortable, as happened at the end of the Permian. Again if it allowed the temperature to reach a torrid 28 degrees Celsius worldwide, then watched it plunge by half to create a highly erratic series of ice ages, as happened between the Eocene (50 million years ago) and the Pleistocene (the start of the ice ages 2.58 million years ago), you might conclude that the instrument was no use. Lovelock even claims that it was Gaia trying to cool the planet which led to the ice ages (page 43, *The Revenge of Gaia*). He never mentions the configuration of the continents, the real cause of the ice ages.

In the years after 1979 Lovelock struggled to gain recognition for his Gaia hypothesis in the academic community, though he received enthusiastic support from some quarters, notably the American biologist Lynn Margulis. In fact Lovelock is rather in the same position as a scientist trying to understand how life got started – he is faced with the end product, so it must have happened, but how exactly? Lovelock deserves credit for explaining the organic origin of the atmosphere, but his Gaia thesis of 1979 is distinctly short of mechanisms to explain how exactly the atmosphere is controlled, and in fact, we still do not understand it.

## Chapter 18 – Recent Geology

In 1980 came the first evidence of the most dramatic event in the last 100 million or more years of the history of the Earth. The subject was the boundary between the Cretaceous period, which ended 65.5 million years ago, and the start of the Tertiary which followed it. The research came from the American geologist Walter Alvarez (1940-). He had the audacity to claim that Cretaceous period ended with an almighty bang – the impact of a meteor, nine kilometres (5.6 miles) across. Few species of fossils pass through this boundary, known as the K–T boundary (Kreide (chalk)-Tertiary). In those places where there is continuous deposition of sedimentary rock through the K-T boundary, the transition certainly appears very sudden.

Alvarez found evidence for this at Gubbio in Italy. Here he found extraordinarily high levels of the element iridium – ten times the expected amount – in a thin layer of clay deposited at the K-T boundary. Iridium is comparatively rare on earth, but is common in meteorites. In 1980 Alvarez published a paper in the American journal *Science*, jointly with his father, Luiz Alvarez, a Nobel prize-winning astronomer (this helped to ensure publication!) Professional geologists were of course sceptical at first, but since publication, the evidence has mounted to the point where it has become overwhelming. Iridium layers at the K-T boundary have been found worldwide – in Tunisia, France, Denmark and Texas for example. From some of these sections, "shocked quartz" has been recovered, indicating its deformation at high pressure. The regular flora apparently disappeared utterly, to be replaced by ferns – the "fern spike" – which are well-known as a pioneer plants because they spread spores so easily. Then in 1991 the smoking gun was finally identified – a crater of an appropriate age, at Chicxulub on the Yucatan peninsula in Mexico, 200 kilometres (125 miles) across.

Despite the accumulating evidence, not all geologists accept that a meteor was responsible for the catastrophe at the end of the Cretaceous, partly because there is another strong candidate. At the same time, in the area that is now India, massive lava beds called the Deccan Traps were erupted. The Traps are on another scale altogether from Hawaiian-style shield volcanoes. 2.5 million cubic kilometres (600,000 cubic miles) of basalt were erupted over a time span of probably only 1.5 million years. This is enough to cover the whole of the area of Alaska and Texas to a depth of one kilometre. In addition, this material was poured out on top of existing continental rock, not onto the ocean floor. This also happened at the eruption of the Siberian Traps, which brought down the curtain on another great extinction at the end of the Permian. These continental traps deny the principles of Lyellism, because nowhere on earth today can this phenomenon be observed. They share their non-uniformitarianism with the Chicxulub meteor, but there can be absolutely no doubting the evidence. It is certainly striking that the two greatest mass extinctions – out of five altogether – have coincided with these truly vast lava eruptions. Also, eruptions of this type bring up unusually large quantities of iridium from the mantle.

There is a third explanation for the disaster at the K-T boundary – a worldwide fall in sea levels, enough to increase the land area of the continents by a quarter. One can't help feeling that this happened because the lava which had been beneath the crust in the Deccan was now on top of it, leaving a large hole to fill in the mantle! This would cause the shallow seas to withdraw from the continental shelf. Also – can it be a coincidence? – exactly this set of circumstances was also found in the greatest extinction of them all, at the end of the Permian, when the Siberian Traps were created and there was also a marine regression.

The effects of blocking out the sun for a period of months or years are not difficult to imagine, but the lack of sunlight was evidently not the only problem. The mean world temperature would have plummeted rapidly, itself accounting for many fatalities. Some writers have postulated torrential rains of sulphuric acid. So the phytoplankton in the sea perished, destroying the base of the food chain; the same thing happened to the green plants on the land. Herbivorous animals, which depended on plants and plankton as their food, died out as their food sources became scarce; in turn, top predators such as *Tyrannosaurus rex* also perished. Foraminifera and molluscs, including ammonites, as well as organisms whose food chain included these shell builders, became extinct or suffered heavy losses. Nothing weighing more than 35 kg (80 lbs) survived. Most famously, the dinosaurs disappeared forever.

\*\*\*\*\*\*\*\*\*\*\*\*\*\*\*\*\*\*\*\*\*\*\*\*\*\*\*\*\*\*\*\*\*\*\*\*\*\*\*\*\*\*\*\*

In recent years there have also been further advances in palaeontology, one of which concerns the very first land animal. At some point in the Devonian (416-360 million years ago), what had previously been a vertebrate fish found its way onto the land and became the first terrestrial vertebrate. Recent genetic evidence showing the similarities between all terrestrial vertebrates indicates that this may only have happened once, so this humble fish became the founder of the tetrapod group which includes all amphibians, reptiles (including the dinosaurs), birds and mammals. The tetrapods ("four feet" in Greek) characteristically have four limbs, evolved from the pectoral and pelvic fins of the ancestor fish. It used to be thought that this was a lobefish, the lobes developing into muscly limbs. Modern opinion, backed by the evidence of cladistics – which takes into account all the attributes of ancestral creatures – favours the lungfish as the ancestor. Species still survive today in muddy Australian pools. After all, whatever it was would have needed lungs on the land! (In addition, the lungfish has a large bone, corresponding to the human upper arm or humerus, which attaches to its "shoulder", though in other respects its fins do not resemble the single bone – two bones – wrist – digits pattern of the later tetrapods.) This momentous event occurred around 397 million years ago. In any event – lobe or lung – modern humans have inherited some rather important features from their fishy ancestors – one central spine, two arms, two legs, jaws, teeth, lungs, and also the practice of both eating and breathing through the mouth.

Another feature of the tetrapods is five-fingeredness. Fossil creatures from the Devonian and Carboniferous have been found with different numbers of fingers, but five is the number which survived. It made the ten times tables easier! Also, as the tetrapods evolved, so did the four-limbed nature of the beast. Snakes are tetrapods which have lost their legs, and birds have used two of them to make wings instead.

One of the problems in elucidating the emergence of tetrapods from fishes has been a great shortage of actual specimens of intermediate species in the fossil record, where there was a gap of about 80 million years between the indisputable amphibians of the Carboniferous and the lung or lobefish which were their nearest known relatives from the Devonian. Over the last 20 years this position has been rectified by the full description of two early amphibians from the late Devonian. One is *Acanthostega* ("thorny covering"), an armoured newt only 7 cm (3 inches) long, which still had internal gills like a fish. The other is *Ichthyostega* ("fishy covering"), about a metre (three feet) or a metre and a half long and without internal gills. *Ichthyostega* was discovered in Greenland in the 1930s. It came to be the exclusive property of a "difficult" Norwegian palaeontologist called Jarvik, under the academic rules of intellectual rights, and so beyond the reach of other paleontologists. The deadlock was eventually broken by a British scientist, Jennifer Clack, who persuaded Jarvik to allow her onto the site in Greenland to study *Acanthostega*, a much smaller amphibian found in the same late Devonian rocks. (Professor Clack, née Agnew, started her career as a zoologist. She knew she had found her man when she met her future husband at a biker's rally and the word "Dimetrodon" cropped up in his conversation.

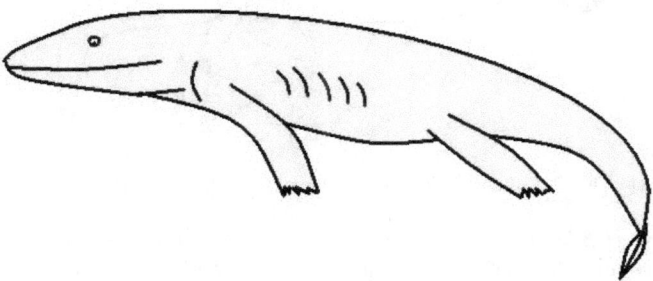

Steggy the Ichthyostega

Clack and her assistants, working on notes left by a Cambridge student from the 1970s, at first despaired of finding anything – a great risk for such costly expeditions. However specimens were eventually located and taken back to Cambridge where they were exposed after many painstaking hours – in fact months – of delicate drilling using dental equipment. This work provided Clack with the paleontologist's dream, the first complete specimens of previously barely known creatures. *Acanthostega* has both limbs and digits at the end of them – in fact eight fingers on each of its forelimbs (she had found genetic freaks, according to the Ibsenesque Jarvik, like six-fingered people). The limbs and digits are not however thought strong enough to support the animal out of water, but must have evolved to support it within water. Such animals must have later evolved the wrists and elbows needed for movement on the land. When finally fully revealed,

*Ichthyostega* turned out to be a larger but similar creature, this time with seven digits on its rear limbs.

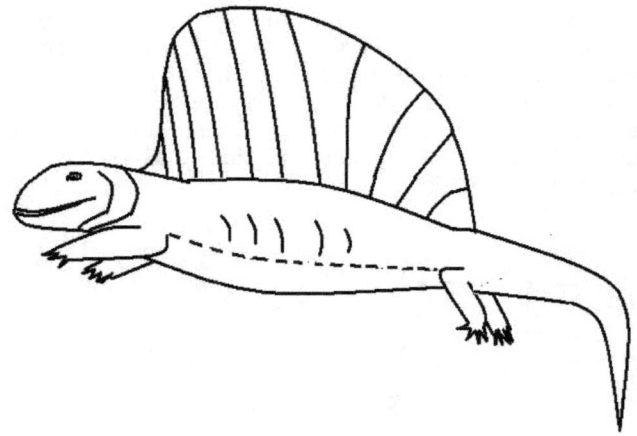

Dimetrodon, a sail-backed reptile from the Permian – in fact a synapsid

Because *Ichthyostega* already possessed fingers and toes it could tell the anatomists little about the origin of limbs. *Acanthostega* proved more interesting because of its flipper-like limbs, but scientists working in the field thought there must be other missing links, expected to date from about 375 million years ago, 10 million years before *Acanthostega*. After four summers of fieldwork in the far north, the American anatomist Neil Shubin and his colleagues identified just such an earlier creature from Devonian beds on Ellesmere Island in the Canadian Arctic. Given the Inuit name *Tiktaalik*, this fish had articulated fins with primitive shoulders, elbows and wrists which would have enabled it to perform the fishy equivalent of press-ups, presumably to get out of the water and away from the large fishy predators which infested the deltas in which it lived. This animal and others like it also developed flexible necks so that they could eat without moving the rest of their body. Fishes do not have this feature – they cannot look over their shoulders. In addition, *Tiktaalik* had a flat head with eyes on the top, like the later amphibians and of course similar to crocodiles and snakes today. Fish, by contrast, have conical heads with eyes at the sides. When first announced to the world press in 2006, Tiktaalik hit the front pages of the New York Times.

**BIBLIOGRAPHY**

Charles Darwin, The Voyage of the Beagle, Penguin Classics (2003) (first published 1839)

Gaia – A New Look at Life on Earth, James Lovelock, Oxford University Press (1995)

The Revenge of Gaia, James Lovelock, Allen Lane (2006)

The Weather Makers, Tim Flannery, Penguin (2005)

H2O A Biography of Water, Philip Ball, Phoenix (1999)

The Beautiful Basics of Science, Natalie Angier, Faber and Faber (2007)

Your Inner Fish, Neil Shubin, Penguin (2009)

Science – A four Thousand Year History, Patricia Fara, Oxford University Press (2009)

Science – A History, John Gribbin, Penguin (2002)

Mendeleyev's Dream, Paul Strathern, Penguin (2000)

The Language of the Genes, Steve Jones, Flamingo (1993)

A Brief History of Times, Stephen Hawking, Bantam (1988)

The Selfish Gene, Richard Dawkins, Oxford University Press (1989)

The Mould in Dr Florey's Coat, Eric Lax, Abacus (2004)

Isaac Newton – The Last Sorcerer, Michael White, Fourth Estate (1997)

Homage to Gaia, James Lovelock, Oxford University Press (2000)

The Age of Wonder, Richard Holmes, Harper Press (2008)

Oxygen – The Molecule that made the World, Nick Lane, Oxford University Press (2003)

A History of Physics, Florian Cajori, Dover (1962)

The Origin of Life, Paul Davies, Penguin (2003)

Wrinkles in Time, George Smoot, Abacus (1993)

The Map that Changed the World, Simon Winchester, Penguin (2001)

Ice Age, John and Mary Gribbin, Allen Lane (2001)

The Strangest Man, Graham Farmelo, Faber & Faber (2009)

Genome, Matt Ridley, Fourth Estate (1999)

A Short History of Nearly Everything, Bill Bryson, Doubleday (2003)

Periodic Tales, Hugh Aldersey-Williams, Penguin (2012)

# INDEX

Absolute zero 73
Academy of Natural History 85
Acanthostega 216
Aeroplane 122
African Eve 168
Agassiz, Louis 82
Age of the Earth (book by Arthur Holmes) 171
Age of the Earth 48, 49,134, 177
Air Chemistry 58
Air pump 35
Alchemy 13
Alexander, Albert 161
Alexandria 9
Alkali Metals 108
Allen, Paul 203
Allosaurus 85
Almagest 14, 15
Alpha Centauri 188
Alpha Ray 133, 135
Alpher, Ralph 193, 195
Altair 8800 203
Alum 79
Aluminium 76
Alvarez, Walter 214
Alvin 185
Amino Acid 158, 163
Ammonia 211
Analytical Engine 200
Andromeda 191
Anion 99

Annalen der Physik 130
Anning, Mary 79
Anti-gravity 198
Apatosaurus 85
Apple II 203
Archaeopteryx 92
Archimedes 11
Argon 109
Aristarchus 11
Aristotle 10, 14
Arithmetic and Logic Unit 201
Arrhenius Svante 143, 209
ASCII 208
Aston, Francis 135, 190
Astronomical Unit 20, 186, 187
Atom 9, 32
Atomic Bomb 198
ATP 167
Avagadro, Amadeo 72
Avagadro's Law 72
Avery, Oswald 152
Avicenna 12
Ayscough., William 37
Babbage, Charles 200
Babbington, Henry 37
Bacteria 119, 120
Barometer 29
Bateson, William 150
Battery 63
Baude, Walter 1987
Beagle 87
Becquerel, Henri 117
Bell Laboratories 194

Benatke Castle 24
Benz, Karl 122
Bergen School of Meteorology 172
Beringia 177
Berzelius, Jonas 73, 74
Bessel, Friedrich 188
Beta Ray 133, 146
Bethe, Hans 195
Big Bang 193
Binary Code 204
Black Body Radiation 129
Black Hole 64
Black Hole 64, 197
Black Smokers 164
Black, Joseph 52
Bletchley Park 201
Bohr, Neils 137
Boltzman, Ludwig 113
Bone Wars 85
Boolean Logic 204
Boolean Operations 205-208
Borlaug, Norman 166, 167
Boson 147
Boulder Clay 83
Boulton, Matthew 54
Boyle, Robert 35, 36
Boyle's Law 35
Brachiosaurus 85
Bragg, Lawrence 139
Bragg, William 139
Brahe, Tycho 21-3
Brand, Henning 35
Brandt, Georg 51

Brief History of Time 197
Brightness of Stars 198
Broglie, Louis de 139, 140
Bromine 76
Brown, Robert 126
Brownian Motion 125
Bruno, Giordano 21
Buckland, William 83
Buffon, Compte de 48
Bunsen, Robert 108
Cadmium 76
Caloric 64
Calvin-Benson Cycle 166
Camarosaurus 86
Cann, Rebecca 168
Cannizzaro, Stanisloa 102
Canon of Medicine 12
Carbolic Acid 120
Carbon 14 137
Carbon Atom 104
Carbon Dioxide 52
Carbon Dioxide 52, 209, 210
Carbon ring 144
Carbon Rings 102
Cartesian Coordinates 32
Cassini Divisions 34
Cassini, Giovani 33
Catastrophism 81
Cathode Ray Tube 114, 115
Cation 99
Cavendish, Henry 56, 57
Cells 149
Cepheid 190

Cerium 74
CERN 147
Chabaneau 75
Chadwick, James 136, 146
Chaim, Ernst 161
Challenger Deep 178
Chambers, Robert 89
Chargaff, Erwin 152, 157
Charles' Law 72
Charles, Jacques 72
Charpentier, Jean de 82
Chemical Bond 142
Chemical Notation 74
Chicxulub 214
Chlorine 66
Chloroplast 164
Chromosome 149, 153
Circulation of the Blood 30
Circumference of the earth 11
Citric Acid Cycle 155
Clack, Jennifer 217
Claus, Karl 75
Clausius, Rudolf 113
CNO Cycle 195
Cobalt 51
COBE 194
Cobol 202
Codon 157
Cold Front 173
Colossus 201
Comet 23
Comma-free Code 158
Computer Architecture 201

Computer Program 200
Constantinople 13
Cope, Edward 85
Copenhagen Interpretation 141
Copernicus 13, 17-20
Copper hemispheres pressure 34
Core Memory 203
Covalent Bond 143
Crab Nebula 196
Cretaceous, End of 214
Crick, Francis 156, 158
Cronstadt, Axel 51
Crookes Tube 116
Crookes, William 114
Crystal Lattice 140
Crystal Sphere 14
Curie, Marie 117
Curie, Pierre 117
Cuvier, Georges 49
Dalton, John 70-72
Darwin, Charles 86-89, 91, 92
Darwin, Erasmus 86
Davy, Humphrey 64-6
Dawkins, Richard 160
De Humani Corpori Fabrica 31
De Nova Stella 21
De Rerum Natura 13
De Revolutionibus Orbium Coelestium 19, 20
Democritus 9
Dephlogisticated Air 56
Descartes, Rene 32
Descent of Man 92
Devonian Period 216

Diamond 144
Dicke, Robert 194
Difference Engine 200
Digges, Sir Thomas 20
Dimethyl Sulphide 210
Dimetrodon 217, 218
Diologo 29
Diplodocus 85
Dirac, Paul 142
DNA 145, 151, 152, 159, 160
Döbereiner, Johan 76
Double Helix 157
Double-slit Light Experiment 69
Drosophilia melanogaster 153
Dynamical Theory of the Electromagnetic Field 110
Earth, Age of 48, 49, 134, 137
Earth, Weight of 58
EBCDIC 208
Eddington, Arthur 131, 189, 196
Edison, Thomas 117
Ehrlich, Paul 121
Einstein, Albert 124-131
Electric charge of elements 73
Electricity, Static 62
Electromagnetism 98, 99, 110
Electron 115, 135
Electron Shell 138-139
Electrons as Waves 139
ENIAC 201
Epicycle 15
Epitome 17
Equant 15

Eratosthenes 11
Essay on the Principle of Population 91
Ether 10, 111
Etudes sur les Glaciers 82
Euclid of Alexandria 11
Eukaryotes 164, 166
Eureka 11
Europium 75
Event Horizon 197
Face of the Earth 171
Fahrenheit, Gabriel 51
Fairchild, Thomas 94
False Geber 12, 13
Faraday, Michael 67, 97-100
Fermi, Enrico 146
Fermion 147
Ferranti Mark 1 202
Feynman, Richard 147
Fitzroy, Robert 87
Flame Test 109
Fleming, Alexander 161
Fleming, Walther 149
Floating Point Numbers 208
Florey, Howard 161
Flowers, Tony 201
Flyer 1 122
Foot-pound 105
Fortran 202
Fourier, Jean 49
Fowler, Willy 196
Frankland, Edward 101
Franklin, Benjamin 62, 63
Franklin, Rosalind 157

Frauenberg 17
Frauenhofer Lines 70
Frauenhofer, Josef von 70
Fresnel, Augustin 70
Fruit Fly 153
Gaia 211
Galapagos 88
Galen 11
Galilei, Galileo 27-29
Galvani, Luigi 63
Gamma Ray 133
Gamow, George 195, 196
Gas combination by weight 71
Gas, 30
Gassendi, Pierre 32
Gates, Bill 203
Gay-Lussac 71, 72
Geber (Jabir) 12
Geiger Counter 135
Geiger, Hans 135
Geissler, Heinrich 114
Gene 150
General Theory of Relativity 130, 131
Geography (Ptolemy) 16
Geological Maps 79
Germanium 105
Gibbs, Willard 113
Gilbert, William 26
Glossopteris 170
Gondwana 170
Goodyear, Charles 122
Graptolites 84
Gravity 38

Great Red Spot 33
Greenhouse Gases 209
Griffiths, Frederick 152
Griffiths, John 157
Halley, Edmund 40, 44, 45
Halogens 76
Harvey, William 31
Hawking, Stephen 197
Heat, Latent 52
Heat, Specific 52
Heatley, Norman 161
Heezen, Bruce 179
Heisenberg, Werner 141
Heliocentric system 14, 18, 19
Helium 109, 133, 190
Helmont, Jan van 30
Henderson, Thomas 188
Herschel, William 64
Hertz, Heinrich 111
Hertzsprung, Enjar 188
Hertzsprung-Russell Diagram 189
Higg's Boson 147
Higgs, Peter 148
Hippocrates 9
Histoire Naturelle 48
History of Plants 46
HMS Challenger 178
Hodgkin, Dorothy 162
Hollerith, Herman 201
Homes, Arthur 171
Hooke, Robert 33, 36
Hoyle, Fred 193
Hubble, Edward 191

Hubble's Law 192
Human Genome Project 160
Hutton, James 77
Huxley, Thomas 92
Huygens, Christian 32
Hybrid plants 94
Hybrid Wheat 166, 167
Hydrogen Bond 145
Hydrogen in Acid 66
Hyperthermophiles 164
IBM 202
Ice Age 175-177
Ichthyosaur, 80
Ichthyostega 216
Iguanodon 79
Illustrations of the Huttonian Theory 78
Immunization 119
Inclined Plane 27
Inert Gases 109
Inflammable Air (hydrogen) 57
Ingenhousz, Jan 64
Inorganic and Organic molecules 73
Internal Combustion Engine 121
Ion 99
Ionic Bond 143
Iridium 214
Iridium 75
Irrational number 9
Isotope 136, 137
Isotopes 104
Jacquard, Joseph-Marie 200
Jansen, Pierre 109
Jaramillo Event 183

Jarvik 217
Jenner, Edward 119
Jobs, Steve 203
Joliot-Curie, Frederic and Irene 136
Joule, James 105, 112
Jurassic Fossils 79, 80
Kekule, Friedrich 102
Kelvin, Lord (William Thomson) 73, 106, 107
Kepler, Johannes 20, 23-25
Kepler's Laws 24-26
Kirchhoff, Gustav 108
Kleist, Edwald 63
Koch Postulates 120
Koch, Robert 120
Krebs Cycle 155
Krebs, Hans 155
Krypton 109
K-T Boundary 214, 215
Lamarck, Jean-Baptiste 87
Language of the Genes 169
Lanthanides 75
Lapworth, Charles 84
Latinised element names 74
Lavoisier, Antoine 59-61
Law of Octaves 102
Laws of Thermodynamics 107
Leavitt, Henrietta 190
Leiden Jar 62
Lemaitre, Georges-Henri 193
Lenoir, Etienne 121
Lepton 147
Leucippius 9
Leucocyte 150

Levene, Phoebus 151
Lewis, Gilbert 143
Life Force 74
Linnaeus, Carl 47
Lister Joseph 120
Lobe Fish 216
Lockyer, Norman 109
Locomotive 67
Logarithms 26
Lorentz Transformation 127
Lorentz, Heinrich 126
Lovelace, Alice 200
Lovelock, James 211
Lowell Observatory 191
Lucretius 13
Lunar Society 53
Lung Fish 216
Lyell, Charles, 80, 81
Macintosh 204
Magnetism 26
Malaria 93
Malpighi, Marcello 32
Malthus, Thomas 91
Mammalia 47
Manhattan Project 198
Mantell, Gideon 79
Margulis, Lynn 165
Marianas Trench 178
Mars Orbit 24
Marsden, Eric 135
Marsh, Othniel 85
Mass Spectrometer 135
Mather, John 194

Matiotte's Law 36
Maxwell, James Clerk 109-113
Maxwell-Boltzman Distribution 113
McClintock, Barbara 167
Mechanism of Mendelian Inheritance 155
Medieval Warm Period 209
Mendel 150
Mendel, Gregor 94-96
Mendeleyev, Dmitri 102-105
Menghini, Vincenzo 32
Methane 211
Michelson, Albert 112
Michelson-Morey Experiment 112
Microbiology 119
Micrographia 36
Microwave Radiation, Background 194
Mid-Atlantic Ridge 178, 179
Miescher, Friedrich 150
Milankovitch Cycles 174
Milankovitch, Mulitin 174
Miller, Stanley 163
Millikan, Robert 130
Minkowski 124, 128
Mitchell, John 58, 64
Mitochondria 164
Mitochondrial DNA 167
Mons Olympus 181
Moon 15
Moore, Gordon 203
Morgan, Thomas 153
Morley, Edward 112
Morrison Formation 84, 85
Mount Etna 81

Mount Wilson Observatory 191
Murchison, Sir Roderick 80, 83, 84
Musschenbroek, Pieter van 62
Napier, John 26
Narratio Prima 19
Natural History Museum, South Kensington 86
Natural Selection 90-93
Negrito 169
Neon 109
Neptune 76
Neptunium 198
Neumann, John von 201
Neutrino 146, 147
Neutron 136
Neutron Star 1987
New System of Chemical Philosophy 71
Newcomen, Thomas 53
Newlands, John 102
Newton and the apple 38
Newton, Isaac 36-43
Newton's Laws of Motion 39, 40
Nickel 51
Nitric Acid 12
Nitrous Oxide 66
Nobel Prize 116
Nobel, Alfred 116
Noble Gases 109
Occam's Razor 81
Oersted, Hans Christian 98
On the Electrodynamics of Moving Bodies 126
On the Nature of Limbs 86
On the Relation of the Properties 103
On the Silurian and Cambrian Systems 84

Ordovician Period 84
Origin of the Continents and Oceans 171
Origin of the Eukaryotic Cells 165
Origin of the Species 90
Osiander, Andreas 19
Osmium 75
Osmosis 125
Owen, Sir Richard 86
Oxygen 55
Palaeomagnetism 178, 183
Palladium 75
Parallax 20, 186, 187
Parsimony 81
Pascal, Blaise 30
Pasteur, Louis 119
Pasteurisation 119
Patterson, Clair 177
Pauli, Wolfgang 146
Pauling, Linus 145
Pauling, Linus 145, 155
PDP-8 203
Peabody Museum 85
Peebles, Jim 194
Pendulum Clock 33
Penicillin 161, 162
Penzias, Arno 194
Period Luminosity 190
Periodic Table 102-105
Peripatetics 27
Perrin, Jean 126
Phlogiston 55
Phosphorus 35
Photoelectric Effect 129

Photon 129
Photosynthesis 101, 266
Photosynthesis 64
Pi 9
Planck, Max 129
Planck's Constant 129
Plant Breeding 95
Plate Tectonics 178
Platinum 75
Playfair, John 78
Plesiosaur, 80
Pluto 191
Plutonium 198
Polio 162
Polypeptide Chain 156
Positron 142
Potassium 66
Precession of the Equinoxes 174
Priestley, Joseph 53-6
Principia Mathematica, Newton 40
Principles of Geology 81, 88
Prokaryotes 164, 166
Proton 135
Proust, Jean 71
Pterodactyl 80
Ptolemy 11, 14, 15, 16
Punched Cards 200
Pythagoras 9
Quanta of Light 129
Quantum Electrodynamics 147
Quark 147
Quasar 193
Rabies, 119

Radio Waves 111
Radioactive Decay 134
Radioactivity 117-118
Radium 118
Ramsay, William 109
Rare Earths 74
Ray, John 40, 46
Red Shift 191
Regiomontanus 17
Resonance 195
Retrograde Motion 15
Rhaeticus 19
Rhodium 75
RNA 151, 159
Roberts, Ed 203
Rochees Moutonees 82
Rocket 68
Roentgen, Wilhelm 116
Rome, Ole 34
Royal Institution 65
Rudolphine Tables 26
Rumford, Count (Thompson) 64, 65
Russell, Henry 189
Ruthenium 75
Rutherford, Ernest 133-135
Sachs, Julius von 101
SAGE System 202
Sal Ammoniac 12
Salk, Jonas 162
Salvarsan 121
Scandium 75
Scheele, Carl 56
Schrödinger, Erwin 141, 156

Schrödinger's Cat 141
Sedgwick, Adam 83, 84
Seed types 46
Seismograph 181
Selenium 74
Selfish Gene 160
Semiconductor Memory 203
Sex Linkage 154
Shapley, Harlow 191
Shubin, Neil 218
Siccar Point 77
Sickle Cell Anaemia 93
Siderio Nuncius 28
Silicon 74
Singularity 197
Skeptical Chymist 35
Slipher, Vesto 192
Small Megallanic Cloud 190
Smallpox 119
Smith, William 78
Smoot, George 194
Soddy, Frederick 134
Sodium 66
Solar Eclipse (1919) 131
Solar System, Size 34
Space-time 128, 131
Special Theory of Relativity 126, 127
Spectroscopy 108, 188
Speed of Light 34, 110
Spirit of Sylvester 30
Spring in the Air 35
Standard Candle 191
Starry Messenger 28

Steady State 193
Steam Engine 53
Stephenson, George 67, 68
Stromeyer, Friedrich 76
Strong Nuclear Force 146
Subduction 181
Suess, Edward 170, 171
Sulphuric Acid 12, 13, 57
Sun Composition 194, 195
Sun, Life of 190
Supernova 191, 196, 197
Syphilis, 121
Systema Natura 47
Telescope 27, 33
Tellurium 56
Tennant, Smithson 75
Tetranucleotide Hypothesis 152
Tetrapods 86, 216
Thalassemia 93
Thales of Miletus 9
Tharp, Marie 179
Theory of the Earth 77
Theory of the Gene 155
Thermodynamics 107
Thermometer 51
Thompson, Benjamin (Count Rumford) 64, 65
Thomson JJ 115, 136
Thomson, George 140
Thomson, William (Lord Kelvin) 73, 106, 107
Thorium 74
Three-axis control 122
Tiktaalik 218
Titan 33

Tombaugh, Clyde 191
Torricelli, Evangelista 29
Traite Elementaire de Chemie 60
Transform Fault 183
Treatise of Electricity and Magnetism 111
Triads 76
Tuberculosis 120
Turing, Alan 201
Tycho's Supernova 21
Uncertainty Principle 141
Unification of Forces 147
Uniformitarianism 81
Universe, Size of 190
Uranium 133, 137
Uranus 64
Urea 76
Urea Cycle 155
Urey, Howard 163
Vacuum 32
Vacuum Tube 114
Valency 101
Van der Waals Force 142 143
Velocity of Molecules 112, 113
Verrier, Urbain le 76
Vesalius, Andreas 31
Vestiges of the Natural History of Creation 89
Virchow, Rudolf 149
Visicalc 203
Vitalism 74
Volta, Alessandro 63
Von Guericke, Otto 34
Voyages of the Beagle 89
Vries, Hugo de 150

Vulcanization 121
Waals, Johannes van der 142
Wallace Line 90, 169
Wallace, Alfred 89, 90
Warm Front 171
Water Chemistry 58
Watson, James 156
Watt, James 53
Watzenrode, Lucas 17
Weak Nuclear Force 146
Wegener, Alfred 171
Weisman, August 149
What is Life 156
Wilberforce, Samuel 92
Wilkins Maurice 156
Wilkinson, John 53
Willughby, Francis 46
Wilson, Robert 194
Wilson, Tuzo 183
Woehler, Friedrich 76
Woese, Carl 164
Wollaston, William 70, 75, 98
Wordstar 203
Work 105
Wright, Orville 122
Wright, Wilbur 122
Xenon 109
X-ray crystal structure 139
X-ray Spectrometer 139
X-rays 116
Y-chromosome 168
Y-Chromosome Adam 168
Yellowstone National Park 164

Young, Thomas 68-70, 125
Ytteby 75
Yttrium 75
Zinc 13
Zircon 185
Zoonamia 87
Zwicky, Fritz 197

www.ingramcontent.com/pod-product-compliance
Lightning Source LLC
Chambersburg PA
CBHW060830170526
45158CB00001B/126